First edition

∞ This paper meets the requirements of ANSI/NISO Z39.48-1992 (Permanence of Paper).
Binding materials have been chosen for durability.
Manufactured in the United States of America.

Library of Congress Cataloging-in-Publication Data
Names: Lopez, Roel R., 1969- editor | Sanders, Jenny C. editor | Cathey, James C. editor | Huddleston, Abigail H. editor | Roosevelt, Simon writer of foreword
Title: The art and science of private land stewardship / edited by Roel R. Lopez, Jenny C. Sanders, James C. Cathey, Abigail H. Huddleston; foreword by Simon Roosevelt.
Other titles: Myrna and David K. Langford books on working lands
Description: First edition. | College Station: Texas A&M University Press, [2026] | Series: Myrna and David K. Langford books on working lands | Includes bibliographical references and index.
Identifiers: LCCN 2026011692 (print) | LCCN 2026011693 (ebook) | ISBN 9781648434235 paperback | ISBN 9781648434242 ebook
Subjects: LCSH: Environmental management—United States | Agricultural conservation—United States—Management | Land use—Environmental aspects—United States | Land tenure—Moral and ethical aspects—United States | Landowners—United States—Attitudes | Environmental responsibility—United States
Classification: LCC GE310 .A78 2026 (print) | LCC GE310 (ebook)
LC record available at https://lccn.loc.gov/2026011692
LC ebook record available at https://lccn.loc.gov/2026011693

Unless noted otherwise, all images are provided by and credited to the editors.

Published by Texas A&M University Press
John H. Lindsey Building, Lewis Street
College Station, TX 77843
www.tamupress.com

This book complies with the EU General Product Safety Regulation (GPSR) (Regulation (EU) 2023/988).
For regulatory inquiries and product safety matters within the EU, contact our authorized representative:
Mare Nostrum Group B.V.
Doelen 72
4831 GR Breda
The Netherlands
Email: gpsr@mare-nostrum.co.uk
Website: https://mngbookshop.co.uk/

This publication has undergone a risk assessment and meets applicable safety standards. Traceability identifiers: ISBN, batch number, and publisher contact details are provided for compliance. No hazardous materials or components are included in this product.

CONTENTS

Foreword
Private Land Stewardship in the 21st Century

The turn of the 21st century is an apt moment to consider what we are doing in America with the abundance of natural wealth we have. These riches—the 2 billion acres of mountains and deserts, forests and plains, lakes, and waterways; in short "the land"—were first thought of as a national priority at the beginning of the last century. Today is an anniversary of sorts and timely to mark changes and needs for change.

Perhaps the biggest change in the history of American conservation is in the ownership of land. When the first naturalists—Bartram, Audubon—scouted the continent, most occupants were Native Americans. As British colonies became American states, a spreading portion became towns, farms, and ranches. With the admission of more states, the disbursement of lands for homesteads and state school trust lands, and the designation of lands for recreation, timber production, and range, the United States of America took its current form.

The result is that 70% of the country is now owned privately in a span of conditions from urban centers, farms, forests, ranches, and undeveloped tracts. Though diverse, this largest share is distinct in kind from public lands—perhaps not least in being the easiest to inhabit when compared to the highest mountain ranges, harshest deserts, and soggiest wetlands.

This book is a rich description of what the private share of America's conservation estate contributes to conservation and what it takes to do so. The information here is necessary for seeing the entirety of American conservation and appraising its legacy and aspirations.

The authors of this book include landowners themselves and experts by experience and professional study and practice. Theirs are the insights and special knowledge of the steward: the person with a particular job of care and support. The landowners in these pages are duty bound to tracts of land. The professionals serve the landowners with whom they work and learn. Their collective insights are fired by the necessary choices of landowning families, the results of those decisions, and the accumulation of this knowledge in families and institutions like the Texas A&M Natural

Resources Institute, Sand County Foundation, and other private and university-based conservation programs.

In presenting the facts, opportunities, and dilemmas of private land conservation, these authors demonstrate the core principles of conservation. First, conservation is a responsibility, which these authors have taken. Connection with the land is the second: They are bound in head and heart to land, by knowledge gathered from experience, acquired by study and by devotion to the land and dependence on each other. Conservation requires this connection.

The third principle is that conservation itself is a discipline of both use and care, which are not alternatives but elements composing the whole. This most basic reality of conservation shows throughout the book. Private land stewardship requires daily clarity on the duty to take and leave—that there is no way to live without meeting human needs from the land somewhere, and to pass on the same or better life there must be limits and replenishment in sync with the features and functions of land.

The demonstration of these principles gives the book a grand scope. All conservation depends on responsibility, connection, and discipline. The landowner's is more directly personal. The same duty calls the urban conservation advocate, backyard gardener, hunter and angler, and wilderness adventurer. Anyone can take responsibility for a place. How persistently and well a conservationist pursues responsibility, connection, and the imperatives of use and care is as much a record of a conservation commitment as is the landowner's mark on the land.

One need not own land to conserve and become a steward. Responsibility is more visibly imperative for a landowner than for the tenant venturing from an apartment but is unavoidable for either who proclaims conservation. Any conservationist can find a job to do as an employee or volunteer. It may be as simple as packing out trash found on a trail or campground.

A connection to land is a bond made of knowing a place and loving it. To connect requires living on or in a place at least part time. Whether making a living by residing there or enhancing your life by visiting, you cannot connect without being there, learning the place, and treasuring it. A hiker on public land can do this, and so can a landowner, to whatever extent they can attain.

The union of use and care is expressly written in American conservation policy but not always grasped in conservation debates. Landowners face it every day but can overlook it. Other conservationists face it in policy questions where it has gotten lost.

When Theodore Roosevelt made conservation a national policy, he said it was as much about development as it was about protection. He was speaking generally about imperatives such as developing western irrigation systems, producing timber, and protecting wildlife refuges, forests, and parks. His contemporaries Gifford Pinchot and John Muir spoke of conservation as "natural resources, transformed for human use" (Pinchot, *Breaking New Ground*) and "beauty as well as bread, places to play in and pray in, where nature may heal and give strength to body and soul" (Muir, *The Yosemite*).

Today the twin motivations remain but have evolved in our minds into alternatives instead of fused elements. This has crippled conservation politics with polarized views on development and protection. Ironically, as they have polarized politically, they have become harder to tell apart practically. Development no longer produces commodities only. We develop campgrounds and trails for beauty. We protect habitats with the heavy equipment needed to manage forests and restore streambanks, and even with chemicals to control invasive species. The traffic of public land users affects land in its way as grazing and drilling do in their way. Despite this obvious need for development to produce Muir-like uses of land, common usage of "conservation" has come to exclude the Pinchot-like uses. The principle that conservation is use and care together needs revival.

From the scope of this book the conservation movement can take important guidance. It would be a mistake to judge one another on conservation principles; instead, we should evaluate the country on its awareness and adoption of these principles. Greater awareness and adoption lead to better results. The truer we are to conservation, the better we are in resolving conservation problems and realizing opportunities. Private land stewardship is a model for how to do this.

As you will see, the problems and opportunities of private land conservation fall into a few long-known categories: how landowners respect the public interest in private land; how the public respects private interests; and how the diverse ownerships of small private tracts can be coordinated across the larger areas that determine the condition of wildlife, water, forests, and range. All of these depend on responsibility, connection, and the discipline of use and care.

If the landowner's responsibility reaches beyond immediate needs into a margin of responsibility for the needs of present and future neighbors and the national landscape, then the line where public responsibility begins must be clear. Today these two responsibilities are defined primarily by regulations and subsidies. Regulations define the minimum responsibilities of landowners, and subsidies define the public interest in asking landowners to do more than they choose on their own. Keeping these joint responsibilities in order is easier when guided by what the landowner knows and loves about the land, what the public knows and cares about the landowner on the land, and the respect for the trade-offs.

These conservation principles are opening a new opportunity in reconciling public and private interests. Some landowners today can earn private profits by meeting public interests. We call them conservation bankers. They purchase land—or limited interests in land, such as conservation easements—to maintain wetlands or wildlife habitat that satisfies state and federal policy goals. They seek out these lands through their connection to them: their expertise in evaluating and improving them. They make "use" of them in their natural state. They earn their living by selling the credit of their work to users of land elsewhere who are converting wetlands or habitat to another use.

Conservation principles are also resolving problems and realizing opportunities across landscapes of many ownerships adjacent to public lands. The Malpai Borderlands Group in Arizona and the Blackfoot Challenge in Montana are prime examples. The Malpai group jointly recognized that their million-acre neighborhood of roughly half-private and half-public land was becoming more difficult to ranch as public land policy evolved and invasive plants turned grasslands to shrublands. The Blackfoot group realized that the continuing effects of historic logging, mining, and grazing were colliding with increasing recreational use of the Blackfoot River, threatening all uses. The Malpai and Blackfoot groups both coordinated work on the land (fire management and habitat restoration projects) and agreements among people (deliberation and policy change). Collaboration such as this is made of the combined responsibilities, connections, and uses of all concerned.

At the heart of this book is a simple idea that will help us better resolve disputes of policy, understand and advise new landowners, and raise new professionals who can help us do these things. The idea is that conservation ultimately depends on the relation of people to each other and the relation of people to land. This is what Aldo Leopold said was most interesting to him. This book is thus about what it takes to have successful conservation in the 21st century.

Simon Roosevelt
Hunter-Conservationist

Preface

Private working lands (family farms, ranches, and forests) account for 60% (more than 1 billion acres) of US land. These lands provide food, fiber, wildlife habitat, recreational opportunities, and economic support to rural communities. Working lands filter the water we drink, purify the air we breathe, and provide cultural and societal benefits. The US Fish and Wildlife Service, for example, often recognizes and acknowledges that the health and sustainability of fish and wildlife populations are in the hands of private landowners.

The critical role that private land stewards play in ensuring the management and sustainability of our nation's natural resources cannot be overstated. Theirs is often a thankless job, and their contributions to society are unknown and unappreciated. Despite this, the individuals who work the land do so for the benefit of us all. Driven by an innate and often spiritual connection to the land on which their business depends, no one has a more profound vested interest in the health and sustainability of our natural resources than private land stewards. As Aldo Leopold stated in *The Farmer as a Conservationist*,

> When land does well for its owner, and the owner does well by his land; when both end up better by reason of their partnership, we have conservation. When one or the other grows poorer, we do not.

State and federal natural resource agencies devote significant resources to programs and initiatives to support private land conservation and management. For generations these agencies and programs have been administered by professionals who grew up on farms and ranches themselves and therefore had an inherent understanding of the landowners they were charged with supporting. However, a shift has occurred, and research suggests that up to 70% of recent graduates of agriculture and natural resource university programs have little to no background on private lands. While they have been trained in biology and ecology, they have little understanding of the motivations and character of private landowners and even less understanding of the day-to-day challenges of working the land. As our population becomes increasingly disconnected from the outdoors, sharing and understanding the values and motivations of private landowners are needed to offer perspective to today's natural resource professionals. Truly, this conduit is imperative to the management and sustainability of rural lands.

The Art and Science of Private Land Stewardship seeks to explore the history, philosophy, practice, and use of science-based information to provide perspective to natural resource professionals; conservation advocates; federal, state, and local decision-makers; and the general public in their work to support the working lands that sustain our country. Understanding the character of private land stewards is essential to creating programs and partnerships that will effectively support the resilience and sustainability of our working lands, and thereby natural resources, into the future.

Roel R. Lopez, Jenny C. Sanders, James C. Cathey,
and Abigail H. Huddleston, Editors

The Art and Science of Private Land Stewardship

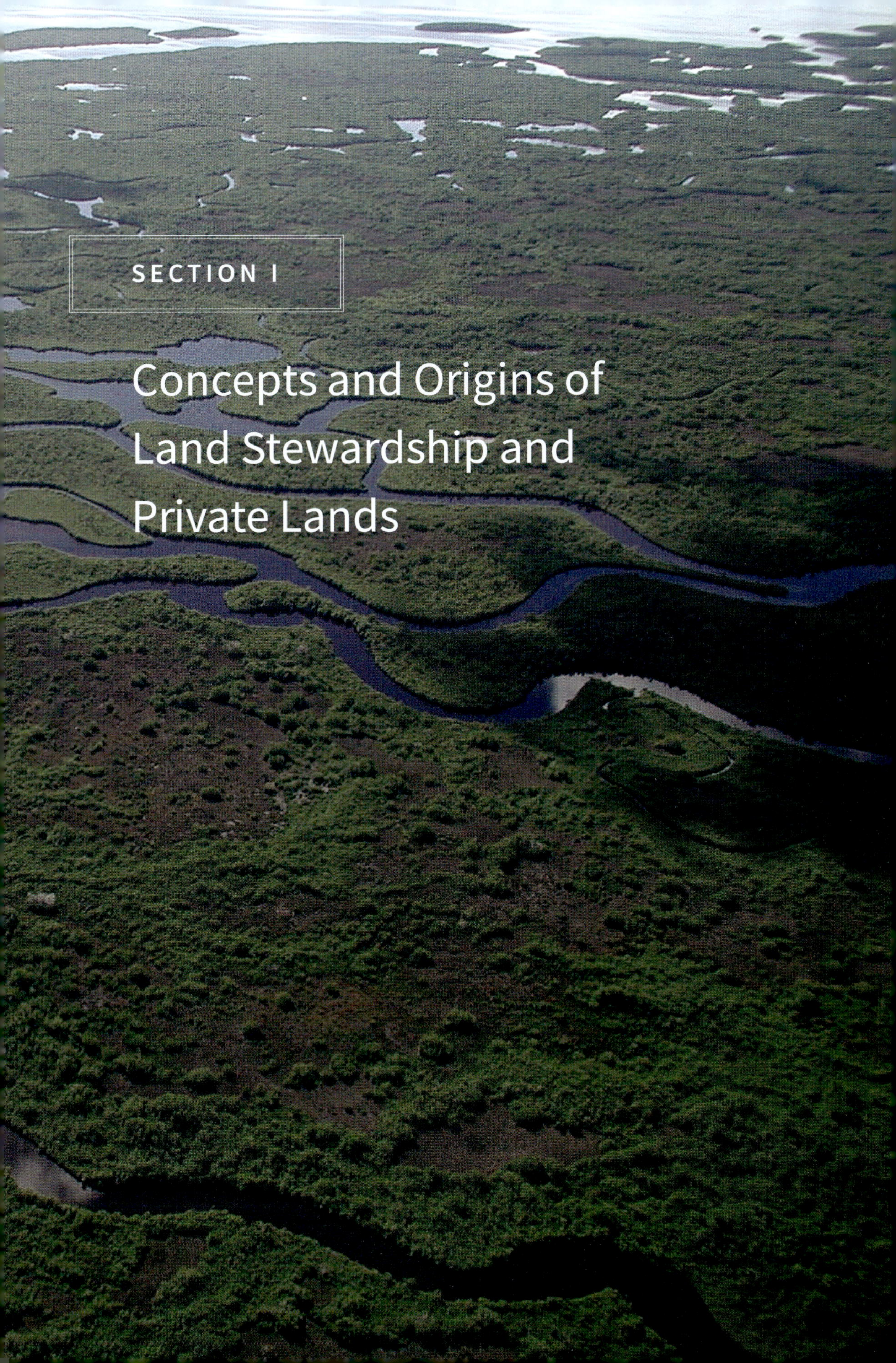

SECTION I

Concepts and Origins of Land Stewardship and Private Lands

Introduction

The topic of land stewardship is an important idea for those engaged in the management and conservation of land-based natural resources. By extension, the products, or benefits of land stewardship, are vital for our future prosperity and quality of life, thus should be of paramount interest to a broader range of policymakers, business leaders, and scientists of all disciplines. While the concept of land stewardship can be applied in some way for all classes of land, it is an increasingly important idea with respect to those lands that are under private ownership. In contrast to public lands, where public policy and agency process prevail in land management decisions, the decisions on private land are mostly driven by individual land managers responding to their own set of designs and incentives. Approximately 60% of the United States' 2.4 billion acres of land area is privately held. What this means is that the bulk of our nation's conservation opportunities arise from private farms, ranches, and forestlands where decision-making is spread among millions of landowners,[1] each enjoying autonomy in their individual pursuits.

The establishment of land as a definable commodity that can be privately owned and traded has a remarkable role in the formation of America. The idea that the United States is inherently unique among nations in a large part draws from a founding concept that land and its outright ownership are fundamental to liberty. Throughout modern history, the relative prosperity of the United States can be attributed to the natural resources from the land—and it is the mixture of exploitation, wise use, conservation, and preservation that has governed our economic success and failures. Most private land was initially in the public domain, and its division and distribution into exclusive ownership were introduced by the nation's Founders with the idea that it was in the public interest and necessary for building a free and democratic republic. Over history, a land ethic has emerged whereby a great many landowners see themselves as not only the beneficiary of their land's productivity but also as the caretaker of a land-based legacy that provides benefits to someone or something beyond themselves. This is what Aldo Leopold meant in 1934 when he posited, "Every landowner is the custodian of two interests, not always identical, the public interest and his own."[2] This is the essence of stewardship.

In this section we delve into both the origins of our nation's private lands and how the concept of private land stewardship developed. Chapter 1 follows the path of development for the unique way in which private lands were established in the United States, while chapter 2 explores how the concept of land stewardship has developed over time in the United States. Together, these chapters lay a foundation for the remainder of this book.

Notes

1. The most recent USDA Census of Agriculture (2017) includes approximately 2 million private farms and ranches, while the US Forest Service National Woodland Owners Survey (2018) estimates approximately 3.7 million private forest owners.
2. Leopold's philosophy is further discussed in the "The Custodian of Two Interests" in chapter 13.

CHAPTER 1

Private Lands
Origins

NEAL WILKINS

In this chapter we review the unique concept of private lands in the United States with a focus on the division and distribution of the public domain. We summarize the origins of private land ownership in the United States and follow the evolution of land ownership from colonization through the postrevolutionary westward expansions of the late 18th and early 19th centuries.

Private Ownership

To fully understand the concept of private land stewardship and its development in the United States, we first start with the idea of what is meant by private land. Private ownership through fee title interest in land is unique among all other private possessions. Land is in limited availability, and its ownership includes the exclusive rights to a variety of naturally occurring resources on a specifically defined geographic location. While land may be owned and controlled by private individuals, corporations, or governments, these owners can neither create nor destroy the existence of this perpetual asset. The rights of land ownership are often referred to with the metaphor "a bundle of sticks," with each stick representing an individual right, or an entitlement to land—for example, possession, control, exclusion, enjoyment, disposition.

Before land could be measured and divided into parcels, it was not generally considered to be private property. While an individual could certainly own the rights to use land, land was not real property until it could be measured and divided into parcels of commonly accepted units of area. The idea that a person could own, trade, or borrow against land as if it were a commodity required innovation. This innovation came in 1538, when Richard Benese published his methods for surveying land and reliably calculating area. As noted by historian Andro Linklater, this was the first time such a process was published in English—making 1538 "the single date when the idea of land as private property can be said to have taken hold" (Linklater 2002, 11).

The notion of property rights was held in high esteem by those who framed the US Constitution. The ideals of many of the Founders, such as Thomas Jefferson, James Madison, and Alexander Hamilton, held that the public interest was best served through establishing strong property rights for individuals. The idea, perhaps radical at the time, was that the transfer of land from the public domain into private hands was considered alone to serve a public benefit. With the Fifth Amendment, the property rights inherent to land ownership were ratified as constitutional rights. This put property rights on par with free speech, freedom of religion, and the right of assembly. James Madison, in his essay "Property," went as far as to state that "government is instituted to protect property of every sort; as well as that which lies in the various rights of individuals" (1792, 266–68).

Division and Distribution

Prior to the American Revolution, land ownership laws in the 13 British colonies followed the English feudal system of Great Britain. All underlying property was held by lords and monarchs, land was leased by a tenant, land was surveyed using a metes and bounds system, and title to land could not be readily transferred. Land was granted under a headright system that was available only to those who could initially pay passage across the Atlantic. In essence, the modern idea of transferable private property in land did not exist for commoners—land ownership, for all but the elite, was generally unavailable.

Those British colonists who could own land had claimed title to property surveyed under the British system of metes and bounds—a system that demarcated land with boundary descriptions that often included movable stones, marked trees, watercourses, and other land-specific monuments. Over time the property corners and boundaries under metes and bounds could become obscure, creating boundary disputes. This practice of surveying was, however, an essential step in the idea of outright ownership of land as property.

The Native Americans who had occupied much of North America did not have the same approach to ownership. Their attitude is best described again by Andro Linklater in his 2002 book, *Measuring America*: "They [Native Americans] shared a pervasive understanding that a particular place belonged to a particular people only to the extent that the people belonged to the place. Rights to the land were gained only by occupation, long usage, or family burial, and they were communal, not individual rights" (43). This was in stark contrast to the ideas of Thomas Jefferson, who believed that possession of land, widely distributed among the citizens of the nation, was the key element to establishing a free and democratic society. Jefferson's vision for creating a nation of yeoman farmers resonated strongly with those who would still be considered commoners in the British homeland, and this especially appealed to more recent Scots-Irish immigrants who were pushing to settle westward of the original colonies. Jefferson's ideas on land ownership and distribution clashed with the push from the ruling British class to limit expansion of land ownership westward from the original colonies. These British limitations on westward land expansion only added to the growing tensions between Britain and the Thirteen Colonies following the French and Indian War (1754–63). With these tensions came a more united American identity, and the demand to own land ultimately contributed to the American Revolution.

Following the American Revolution, our nation's Founders took action to establish a standard means for surveying and distributing land from the territory acquired from Great Britain. In 1785, with language drafted by Thomas Jefferson, the US Congress established a Land Ordinance under the Articles of Confederation. This was ultimately how vast areas of US territory would be put into private ownership. The public land survey system included in the Land Ordinance required division of land into rectangular units of townships, each with 36 numbered sections of 640 acres each. The public land survey system contrasted sharply with the often-complex British system of metes and bounds used by the original colonies, along with Tennessee, Kentucky, West Virginia, and Maine. By efficiently dividing the nation's unsettled territories into definable parcels, Thomas Jefferson's idea was to distribute land to farmer-landowners. He thought that a nation built on private land ownership would advance economic and political independence. Historians continue to speculate on the motivations behind Jefferson's push for the ideal of an "agrarian democracy" in the development of our nation (Malone 1993). Whatever Jefferson's motivation, his actions laid the foundation for future US land policy through the innovations of uniformity in the public land survey system.

Territorial Expansion

By establishing the public land survey system, along with the companion Ordinances of 1784 and 1787, the US Congress introduced the means for land distribution and settlement of the "Northwest Territory"—the land west of Pennsylvania, east of the Mississippi River, south of the Great Lakes, and north of the Ohio River. Upon being surveyed, the land in the Northwest region was sold in parcels of no less than 640 acres (one square mile, also known as a section) and for a price no less than one dollar per acre. Land from these surveys was also reserved for Revolutionary War veterans and for support of public education. These land sales not only generated revenue for a young US government, but they also put land—and title to land—in the hands of settlers. Through this division and orderly transfer of land for settlement, the states of Ohio, Illinois, Michigan, and Wisconsin were eventually established. These states were admitted to the Union by virtue of reaching a threshold of population of settlers; and this settlement was driven by the division, distribution, and privatization of the public domain. Through the process established by the Ordinance of 1787, several civil rights were first protected—one of which was the right to compensation when private property is taken for public use. In 1789, this was further strengthened as the "Takings Clause" included in the Fifth Amendment to the US Constitution. The Fifth Amendment thus further secured title for all US private lands by prohibiting the taking of private property for public use without just compensation.

When the US Senate ratified an 1803 treaty with France, the Louisiana Purchase added 828,000 square miles to US boundaries—effectively doubling the size of the US territories. Following the War of 1812, President James Monroe ordered the survey of the Louisiana Purchase. Starting from the initial point in what is now Arkansas, the new territory was divided and awarded as compensation to war veterans—and then sold to settlers and land speculators.

The Louisiana Purchase was followed, over the succeeding 45 years, by several major territorial expansions. The independent Republic of Texas was annexed in 1845, adding 389,000 square miles of territory. The 1846 Oregon Treaty with Great Britain expanded boundaries by over 288,000 square miles. The 1848 Mexican Cessions added 522,000 square miles, eventually forming California, Nevada, Utah, and parts of Arizona, New Mexico, and Colorado. In 1867, for the price of $7.2 million, the United States acquired an additional 591,000 square miles of Alaska territory from

Russia. In total, from 1781 to 1867, the United States added 1.84 billion acres of territory to the public domain (US Department of the Interior 2020).

Homesteading

With such extensive new territory, there was mounting pressure to accommodate settlement claims to land in the nation's westward expansion. Some of this new territory arrived with illustrations that seemed to suggest a uniform solution. The Republic of Texas, for example, had made land available to citizens at the time of declaring independence. From 1836 to 1842, a head of a household could claim a league (4,428 acres) and a labor (177 acres) totaling 4605.5 acres per family (those arriving in the state after March 2, 1836, were eligible for lesser amounts). From this example in Texas, Andrew Johnson made the first attempt at introducing a Homestead Act to Congress in 1846 (Sioussat 1920).

The Homestead Act eventually passed through Congress in 1862 and allowed settlers to claim title to 160 acres of public land, providing they occupied and cultivated the land for five years. The fact that 160 acres was one-quarter of a 640-acre section made easy use of the public land survey system established by the Ordinance of 1785. The public land survey system was thus used across the broad expanse of most of the western territories of the United States. There were a handful of exceptions, including Texas retaining over 225 million acres of its original public lands, intended to be a chief source of public revenue. In 1850, Texas transferred 67 million acres to the United States, forming parts of the public domain in New Mexico, Kansas, Colorado, and Wyoming.

Through the Homestead Act, the United States eventually transferred about 287 million acres of the public domain into private ownership (US Department of the Interior 2020). Land granted to homesteaders represents the largest single source of "disposition of the public domain" and accounts for 22% of what was once public domain, as well as 12% of the total land area of the United States. The influence of the Homestead Act was multiplied because it was a primary driver for getting the standardized divisions of the Public Land Survey System in place throughout most of the United States. The Public Land Survey System was then key to the private land transfers that included 131 million acres granted to railroads, 61 million acres granted to veterans, and 34 million acres of private land claims (US Department of the Interior 2020). All told, about 1.29 billion acres, or 71%, of the original public domain has been transferred to private or state ownership.

From Settlement to Stewardship

The previous discussion gives a general overview of the acquisition and disposition of the public domain. This is how most of our nation's land ended up in private ownership. What it omits are the frontier battles, wars, displacement of Native Americans, and hundreds of circumstantial details—both good and bad—that add historic context. While these are important details, the point to be made here is that the fabric of our current land ownership pattern was forged during a relatively short period in the nation's history. In the first half of the 1800s, the US territories expanded by more than 30 million acres per year—this is roughly the same as adding land area the size of Virginia to the country's territory every year. This coincided with our nation's largest rate of population growth—population expanded from 5.3 million in 1800 to 23.1 million by 1850.

Historians have traditionally explained this era of US territorial expansion and settlement as being driven by "Manifest Destiny"—a uniform conviction to the idea that territorial expansion was an inevitable and preordained movement for spreading democracy. There are many historians who now are reluctant to use the convenient unifying theme of Manifest Destiny to explain territorial expansion. This alternative view is best stated by historians Andrew Isenberg and Thomas Richards: "We must understand that U.S. westward expansion was neither manifest, nor destiny, but a product of the contingencies of war, diplomacy, demography, economy, and culture—in other words, a product of history" (2017, 4). The policies for division and distribution of the public domain into private ownership were, at the time, in the public interest for absorbing immigration, building an economy, and establishing and defending a nation.

The innovation of the Public Land Survey System when combined with the philosophies of early political leaders promoted a wide distribution of private land ownership that continues to be reflected in the current distribution of private land. The repeated theme throughout the expansion and disposition of US territories was that the distribution of land into private ownership served a public interest. But what does this mean for the future?

The United States' population has expanded over 10-fold in the 170 years since the close of territorial expansion. With a current population exceeding 331 million, most of the US population has shifted to urban centers where they are increasingly disconnected from the land. As our nation has expanded, the sustained productivity of farms, ranches, and forestlands remains as important as ever. By and large, our nation no longer views land-based natural resources as a seemingly unlimited resource. Increasingly it is the conservation value of private lands that is now valued as a public benefit, a concept addressed in chapter 3.

CHAPTER 2

Land Stewardship

A Concept

NEAL WILKINS

In this chapter we review the global idea of land stewardship. For its application to private lands, we focus discussion on the United States with some comparison to British and European concepts, both past and present.

> *Each generation has its own rendezvous with the land, for despite our fee titles and claims of ownership, we are all brief tenants on this planet. By choice, or by default, we will carve out a land legacy for our heirs.*
> —**Stewart Udall**, Secretary of the Interior, 1961–69

In his 1963 book, *The Quiet Crisis*, Secretary Udall expresses the essence of an idea that helped spawn a modern environmental movement. Stewart Udall's call for his generation to act on a nationwide "land conscience" was met with an exceptional response, especially in the US Congress. In the three years following publication of his best-selling book, Congress passed the Clean Air Act (1963), the Wilderness Act (1964), the Water Quality Act (1965), and the Endangered Species Preservation Act (1966). These landmark pieces of legislation are the cornerstones of what is now known as the modern environmental movement. The seeds for this movement can be traced back to the first decade of the 20th century.

In 1908, President Theodore Roosevelt gave an opening address to the Governor's Conference on the Conservation of Natural Resources. Some of his words:

> Nature has supplied to us in the United States, and still supplies to us, more kinds of resources in a more lavish degree than has ever been the case at any other time or with any other people. Our position in the world has been attained by the extent and thoroughness of the control we have achieved over nature; but we are more, and not less, dependent upon what she furnishes than at any previous time of history since the

> days of primitive man. . . . It is safe to say that the prosperity of our people depends directly on the energy and intelligence with which our natural resources are used. It is equally clear that these resources are the final basis of national power and perpetuity. (Roosevelt 1909, 6)

Some 60 years prior to the environmental movement spawned by Udall and his contemporaries, President Theodore Roosevelt inspired a national effort for conserving our nation's public land. Under Roosevelt's leadership, over 230 million acres were set aside as public lands. These national forests, wildlife reserves, national parks, and national monuments remain as some of the best examples of our nation's 620 million acres of public lands.

In the shadow of landmark executive action by Roosevelt and the legislation spawned by Udall, one important idea was obscured, but it was inherent in both President Roosevelt's and Secretary Udall's declarations. This straightforward two-part idea is that (1) the land's natural resources are the basis of future prosperity, and (2) the responsibility for conserving our nation's collective natural resources is largely in the hands of those who own and manage land. In other words, the legacy of land, water, and wildlife left to future generations depends on stewardship by our nation's ranchers, farmers, and foresters.

The land reserves initiated in Theodore Roosevelt's era, combined with the legislative actions of Udall's era, continue to provide an important framework for conserving our nation's land, water, and wildlife resources. Efforts to make progress for conservation by replicating the work of Roosevelt and Udall, however, are not likely to be met with the same magnitude of impact. Because most of our nation's lands are privately owned, it follows that the conservation decisions impacting most of our nation's wealth of natural resources are solidly in private hands. The idea of individual decisions as a driving force for conservation is not to depreciate the importance of implementing public policy via regulation but rather to underscore the overriding importance of those on-the-ground decisions that cannot be easily regulated. For a practical matter, private landowners and their managers now make most of the important decisions that influence the conservation of our nation's natural resources. It is this decision-making responsibility and the subsequent implications for the future of conservation that have driven the modern notion of private land stewardship.

What Is Land Stewardship?

The essence of stewardship is custodial care over a resource of value on behalf of another. This relationship involves three important roles: a steward, an owner, and a beneficiary. Often, the owner and the beneficiary are the same, and the resource is managed by a trustee who is duty bound to the best interests of the owner-beneficiary. A banker who has the custodial responsibility over deposits of an owner-beneficiary is a simple example of such a relationship. This relationship has threads throughout history. In biblical scripture, for example, Jesus uses the "parable of the talents" to illustrate the relationship of a good steward that invests his master's property on his behalf (Matthew 25:14–30; Luke 19:11–27).

While a specific use of the expression *stewardship* as it applies to land emerged only in modern history, the concept of *land stewardship* appears in the philosophy and religious scriptures that reach as far back as civilization itself. The concept of land stewardship seems to have been well understood by many Indigenous cultures. Prior to 1492, the Indigenous population of what is now the United States—which some estimate at roughly 7 million people (Thornton 2005)—occupied and held territories for which they exercised communal property rights. These Native American societies generally held, and defended, such territories for subsistence—agriculture, hunting, fishing, and foraging. Tribal elders would often assign places of hunting, fishing, or foraging to individuals, creating a form of temporary individual property right (Greer 2012). Several Native American tribes have traditional beliefs and values based on a deep reverence for the land along with a commitment to perpetuate the land, water, and wildlife for future generations (Reynolds 2007). This, of course, indicates that a concept of land stewardship existed far deeper into history than the use of the term itself.

Origination of "Land Stewardship"

If the concept of land stewardship predates the use of the term itself, what has prompted the recent use of the term, and why is it important? By 1800, the term *land steward* was in common use in British and American literature, but it was not until after 1940 that the term *land stewardship* was in common use in the same body of literature, and it was not until the 1980s that the term was widely used (figure 2.1).

In British literature, the term *land steward* continues to be used more than *land stewardship*, while in American literature *land stewardship* is in more common usage. It appears that the term *land steward* has its origin in Great Britain, while *land stewardship* is mostly an American term.

Land stewardship requires land stewards, and throughout most of history the landowner or a tenant has served all the functions of a land steward. It took the changes in British land ownership and tenure resulting from

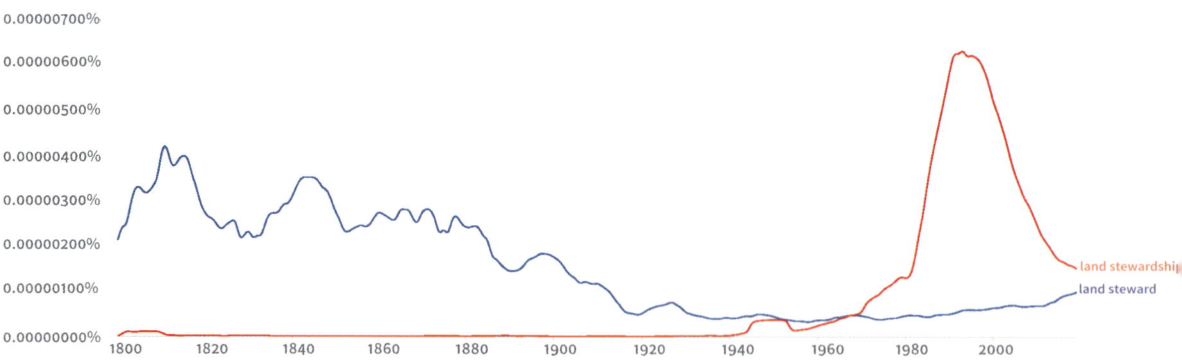

FIGURE 2.1. Comparison of the use in literature of the term *land steward* and *land stewardship* (from Google Books Ngram Viewer for the English language).

the conditions of an agricultural revolution for there to be the demand for a professional land steward. The modern American expression *land stewardship* has obvious ties to the description of one as a *land steward*, but the latter term reaches back at least to the British Agricultural Revolution that began in the mid-18th century. The British Agricultural Revolution, also known as the Second Agricultural Revolution, was marked by an exceptional increase in agricultural output brought on by improved technology and transportation. During the same era, British lands farmed under traditional common rights were transferred into exclusive ownerships, effectively privatizing much of the rural British countryside. Estates in the British countryside could generate substantial wealth from farming and livestock production as a result of new technologies that allowed them to vastly increase production. Mechanization, irrigation, better transportation, advances in animal breeding, and improvements in crop rotation all combined to create a rapid change in land ownership and management. The wealthy owners of these estates needed managers to look after their interests and manage the productivity of the land. It was during this period that the British profession of "land steward" emerged. An early description of the British model of a land steward is described in a 1731 British text by Edward Laurence. This was later summarized by the Reverend John Wilson in the 1847 *Rural Encyclopedia of London*:

> Every land steward ought to possess skill in accounts, acquaintances with the law of landlord and tenant, considerable knowledge of human nature, large experience in negotiating with equals and inferiors, perfect integrity of moral character, much suavity and

> firmness of temper, and a full knowledge of practical agriculture, of land surveying, of planting, and of all the portions of science and mechanics which are connected with the explanation of agricultural phenomenon, and the promotion of agricultural improvement. (67)

Laurence and Wilson's description of a British land steward represents an impressive set of talents that when applied to an estate would benefit the owner of the estate. What this picture of a British land steward lacks is a notion of any benefits except those that would advantage the owner of the estate. This differs greatly from our modern idea of a land steward—an idea that evolved alongside the conditions of American land ownership. In contrast to the British model, the concept of an American land steward evolved from the idea that one entrusted with the private ownership of land had not only a purpose of deriving individual privilege and benefit from the land; the American land steward also had a purpose that was larger than self-benefit. In summary, the origin of the modern concept of land stewardship has American roots, despite the longer history of the British term *land steward*. In fact, as is illustrated throughout the following chapters, the American concept of land stewardship has more in common with the Native American reverence for the land and the belief that natural resources ought to be perpetuated for future generations. The American idea of one being a land steward is more apt to have come directly from this concept of land stewardship than an idea imported from Great Britain at the time of European colonization of North America.

The Public Trust Doctrine: A Case in Point

The management and conservation of wildlife (and wildlife habitats) present one of the clearest cases where private land stewardship is key to the perpetuation of a natural resource for beneficiaries other than the landowner. In North America, wildlife is well established as common property under the public trust doctrine. This principle holds that all wildlife is held by the State to benefit the public—that is, all people, including present and future generations. This is the lynchpin of the North American Model of Wildlife Conservation, and it is what distinguishes wildlife conservation in the United States and Canada from that prevailing in the Old World. In this model, the state or federal government is the trustee, and designated government agencies are obliged to apply their authority for public benefit. In most cases, on private lands, the government cannot compel conservation activities likely to result in the restoration and proper management of wildlife habitats. Agencies and nongovernment organizations may incentivize conservation practices on private lands, but application requires voluntary efforts from private land stewards. Landowners often accept compensation for their wildlife conservation efforts, and this stimulates further investments in conservation practices. This further illustrates the mix of public and private benefits derived from private lands and their stewardship. Commitment to voluntary action implicates private land stewards as caretakers of the fish and wildlife habitat from which the public ultimately benefits. Without such voluntary stewardship the North American Model of Wildlife Conservation would not have succeeded to the degree it has.

In conclusion, although the use of the term *private land stewardship* is often considered a more modern conservation concept, our heritage as Americans is deeply threaded with the notion of the importance of the private citizen in taking care of and wise use of the land and its resources for the public interest. The subsequent chapters in this book take an in-depth look at concepts such as the North American Model of Wildlife Conservation, the evolution of the role of land stewards in conserving our natural resources, the management implications related to trends in land ownership and demographics, the emerging markets and tools for conservation, and the motivations and responsibilities related to stewarding the land and its resources—or what our Founders believed to be the essence of prosperity of our nation and its people.

CHAPTER 3

Public Benefits of Private Lands

CHAD ELLIS, JEFF GOODWIN, AND MYRIAH JOHNSON

This chapter introduces the concept of ecosystem services and illustrates how the impact of private land stewardship extends far beyond the fence lines of individually owned parcels. We explore emerging markets for ecosystem services and look at opportunities for private land stewards to monetize these services to support future investment in the land.

Introduction

Privately owned farms, ranches, and forests, collectively known as *working lands*, are the cornerstones on which our nation was built. These lands not only produce much-needed food and fiber while sustaining rural economies, but they also provide many conservation benefits including clean water, wildlife habitat, and ecological diversity. Collectively, these benefits that humans receive are known as *ecosystem services*. The scientific body of work supporting ecosystem services continues to grow. Research related to soil health, carbon sequestration, wildlife habitat, and water resources shows a direct relationship between healthy open space and improved environmental outcomes for humans.

Almost a century ago, Aldo Leopold recognized the crucial role private landowners would play in conserving our nation's natural resources: "Conservation will ultimately boil down to rewarding the private landowner who conserves the public interest" (1934, 544). While providing critical food and fiber to the nation, working lands also provide societal benefits that are many times overlooked. Well-managed working lands sequester and store carbon, provide wildlife habitats, filter water, support pollinators, and much more. Ecosystem services produce both local and public benefits. For example, a rangeland system supporting higher water infiltration will often have less erosion and produce more herbaceous plant material. This outcome results in direct, on-ranch (or "in situ") benefits to producers, such as enhanced forage availability and quality for livestock. Additionally, water infiltration on rangelands produces off-ranch ("ex situ") benefits for the public by reducing flood and erosion risk in downstream areas.

Simultaneous to this understanding of the benefits of privately held working lands, consumers are increasingly demanding that the companies they do business with demonstrate a commitment to environmental sustainability in the production and distribution of goods and services. This demand has given rise to reporting standards such as environmental, social, and governance (ESG), whereby corporations are judged by their shareholders on their ability to meet specific predesignated metrics that demonstrate their commitment to sustainability. In 2015, the United Nations (UN) issued an urgent call to action, identifying 17 Sustainable Development Goals (SDGs) to be achieved by all UN nations by the year 2030.

While many of the UN SDGs are focused on global poverty and social issues, goals 3, 6, 7, 9, 12, and 13 each have a direct or indirect nexus to nature-based solutions on private lands (figure 3.1). Emerging markets create systems whereby corporations can purchase credits to offset their impacts in these various areas. Credits can come in many forms, with new markets increasingly in development. Common examples include carbon sequestered by well-managed grasslands, crops or forests purchased as

FIGURE 3.1. Seventeen sustainable development goals established by the United Nations Sustainability Council to be achieved by all UN nations by 2030.

offsets for industrial emissions, and wetland mitigation credits for industrial, commercial, or residential development.

The wetland mitigation marketplace is driven by regulatory pressure, but most other markets for ecosystem services are voluntary in nature and therefore are largely being driven by innovation and social pressures.

It is the marriage of corporate demand for sustainability with the realization of the value of privately owned, well-managed working lands that continues to lead to innovations in the ecosystem service marketplace and the focus on nature-based solutions to climate change.

Demand for Ecosystem Services

In 2018, the Noble Research Institute hired Informa Agribusiness Consulting to conduct an economic assessment of the potential value of ecosystem services sourced from privately owned working lands. According to the report, working lands have the potential to supply carbon, water quality, and water quantity credits far exceeding the corporate demand in the same sectors (figure 3.2).

The report estimates a potential demand from the food and beverage sectors alone at $5.2 billion in carbon, $4.8 billion in nitrogen, and $3.9 billion in phosphorus (the latter two as water quality credits) for a total combined value of $13.9 billion that could be contracted with landowners to supply the needed offsets and reward them for the good land stewardship they are implementing.

Previously, when this study was completed, it was assumed that ecosystem services would be monetized through the offset market. That is, the ecosystem service would be marketed separately from any other commodity that was also produced on that land. While offset markets continue to grow, the idea of insetting has grown in prominence in recent years. Insetting is much like offsetting except that it is focused within a corporate entity's own supply chain. With insetting, the ecosystem service is tied to the commodity produced on the land; for example, beef cattle are marketed with the ecosystem services produced. Insetting provides a pathway for improved accounting and transparency in the supply chain, enabling products such as "low-carbon beef." With insetting, corporate entities work within their supply chain to meet their demand for ecosystem services. Additionally, insetting allows ecosystem service credits to remain in the commodity supply chain, enabling those commodities to take credit for the improvements in their environmental footprints and meet climate goals, commonly represented as "carbon neutral by 2050." Selling offsets,

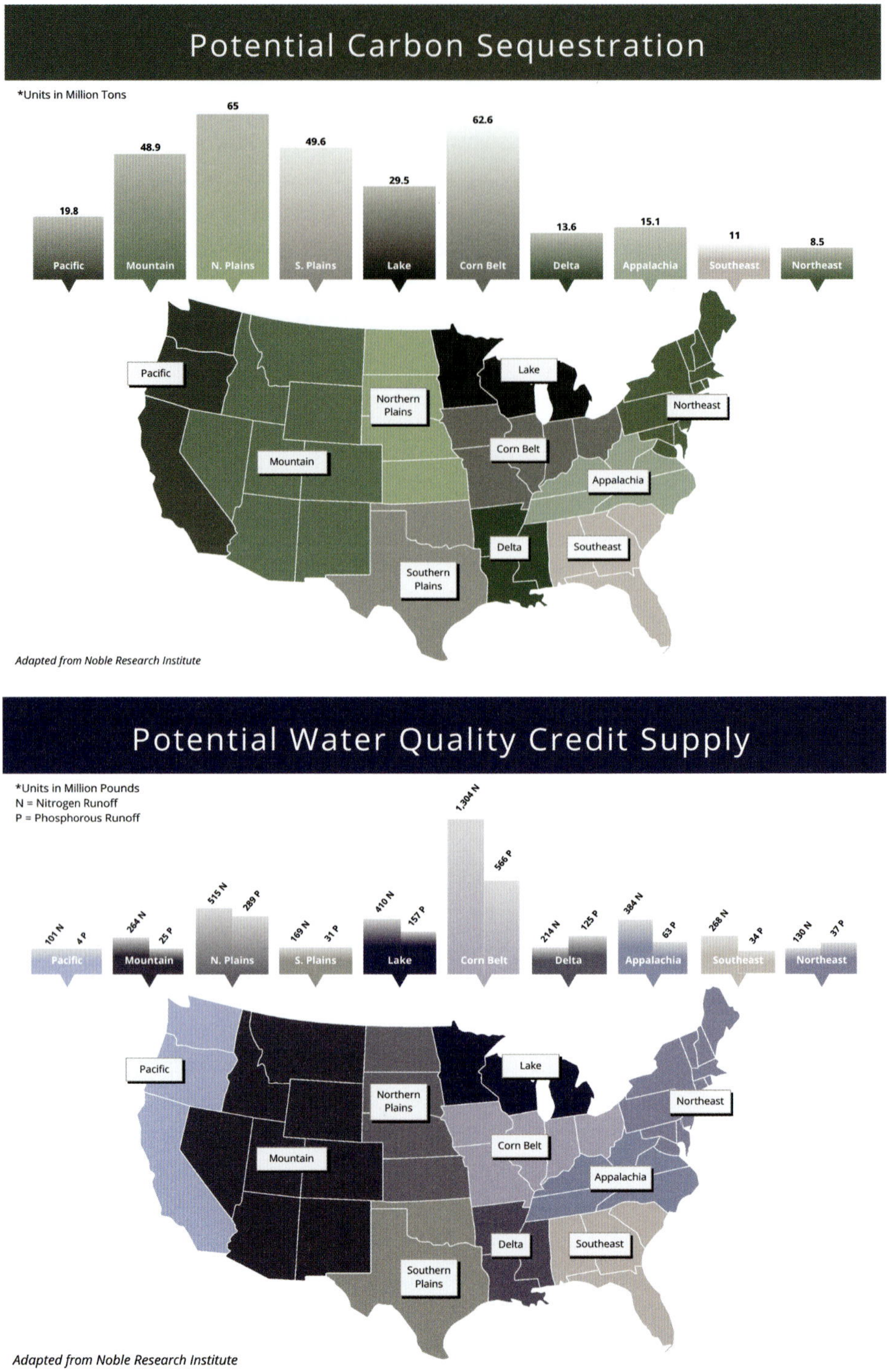

FIGURE 3.2. Working lands across the United States have the potential to supply the ecosystem services for all current corporate demand for credits for those services.

on the other hand, does not allow agricultural supply chains to take credit for those improvements. Rather, the offsets truly do "offset" the impacts of another industry, allowing them to take credit and meet their climate goals. As markets for ecosystem services continue to evolve, there will be debate on which pathway, offsets or insets, should be pursued.

CASE STUDY 3.1

Texas Municipal Water Collaboration with Private Land Conservation

In Texas, regional and municipal water districts are collaborating with the corporate sector to compensate private landowners for conservation practices that have positive downstream impacts on water quality and quantity. Mill Creek, which runs through the Price family's 77 Ranch in Navarro County, Texas, feeds into Richland Chambers Reservoir, a major water source for about 1.8 million Fort Worth residents (figure 3.3). The Price family and others in the area have received incentive payments from the Tarrant Regional Water District and MillerCoors Brewing Company for conservation practices that reduce runoff and erosion. From 2012 to 2016, MillerCoors alone provided $385,000 in incentive payments and technical assistance to help landowners manage land for resiliency.

FIGURE 3.3. Gary and Gary Lee Price gathering cattle along Mill Creek (photo courtesy of Brian Baxter and National Cattleman's Beef Association).

Ecosystem Services of Private Lands

Ecosystem services can be broadly placed in four categories (figure 3.4):

1. Provisioning (food, fresh water, fiber, fuel, etc.)
2. Supporting (cycling of nutrients, soil building, preservation, fertility renewal, photosynthesis, etc.)
3. Regulating (climate stability, moderation of weather extremes, pest control, air and water purification, etc.)
4. Cultural (support of spiritual and cultural heritage, educational, aesthetic, and recreational opportunities, etc.)

The reality is that most private working lands provide benefits in all or most of these categories. Cotton farms produce fiber to clothe families, produce seed that can be used in rations for livestock, sequester carbon, and contribute to agricultural and rural economies. Rangelands produce a renewable dietary source of forage (cellulose) in biomass from a diverse set of grasses, forbs, and woody plants. This cellulose is uniquely converted by ruminant livestock into animal protein and fiber production (e.g., wool, mohair) for human consumption.

Working lands help cycle nutrients, build soil, preserve habitats, and support other services that promote healthy ecosystems. Forest ecosystems provide value through tree roots that anchor nutrients in the soil and prevent runoff and erosion. Forests and rangelands actively managed with prescribed fire support diverse plant communities that improve soil health and support wildlife communities. Farmers plant pollinator-friendly plants

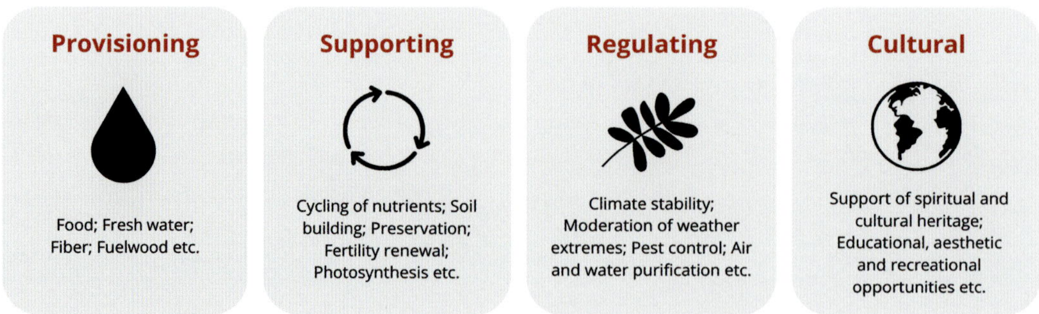

FIGURE 3.4. Recognized broad categories of ecosystem services (graphic courtesy of the National Cattlemen's Beef Association Beef Checkoff Sustainability Program).

along buffers near water bodies, fallow fields, conservation acres, field and road barriers, ditches, and wetlands. The honeybee and monarch populations benefit from the additional source of food and nutrients, and farmers benefit from the healthy soil and water the habitats promote.

While these broad categories of ecosystem services are helpful for speaking the language of the sustainability world, they do not necessarily resonate with the lay person or average producer. In the following sections, we consider five topic areas that encompass many (but not all) working land ecosystem services, threats specific to these services, the role of stewardship within each area, and potential opportunities for producers and managers to derive benefits from ecosystem services in each topic area. Topic areas include food and fiber, water, carbon, biodiversity, and wildlife habitat.*

Food and Fiber

As mentioned previously, working lands produce a renewable dietary source of forage (cellulose) in biomass from a diverse set of grasses, forbs, and woody plants. This cellulose is converted by livestock and wildlife into animal protein and fiber production (e.g., wool, mohair) used for textile production, human consumption, and a host of other human benefits. Ruminant livestock and wildlife (cattle, bison, sheep, goats, deer, and elk), by hosting specialized microbes in their digestive system, can convert cellulose in plant biomass, ultimately producing animal protein and fiber. Livestock and fiber production from working lands serves as a provisioning service that provides economic and/or subsistence value for individuals and local communities. For example, 31.3 million beef cows on 729,000 operations contribute to the $77 billion annual sales for the beef cattle industry (USDA NASS 2019, 2020). As global human populations are predicted to increase exponentially, the sustainable production of animal protein and fiber for human use and consumption will be needed globally (Greenwood 2021).

Opportunities for land stewards.

Relative to other ecosystem services, opportunities for producers and managers to benefit from the production of food and fiber are widely available in well-developed markets for many countries due to existing livestock markets, including local, national, and international trade. Although supported to some degree through private and/or government conservation programs, the current and potential land stewardship ecosystem service co-benefits provided by working land operations to society may not be adequately recognized relative to the societal value this stewardship provides.

---

*The following discussion of the five topic areas is adapted from Society for Range Management (2023).

Water

Water is the dominant land cover type on earth; however, rangelands account for over half of the land surface area, and most of these areas occur in water-limited environments (Millennium Ecosystem Assessment 2005). Water scarcity is often a limiting factor on rangelands, especially those west of the 100th meridian, thus impacting the plants and animals they support and the humans who work and live in these ecosystems. The scarcity of water in rangeland systems and the contribution to the terrestrial land base emphasize its importance. Rangeland managers often work to employ management practices that increase infiltration and reduce runoff to capture every raindrop where it falls, maximizing the amount of water that is available to plants. Maximizing water availability to plants supports many ecosystem services on rangelands because the majority are closely linked to vegetation (e.g., forage for both wild and domestic herbivores, habitat for a wide range of animals, biodiversity, carbon dynamics/sequestration, and nutrient cycling). Some experts have postulated that water is the primary ecosystem service that underpins or facilitates all other ecosystem services (Cotes et al. 2013). Even though water is scarce on many rangelands, the quality and quantity of clean water are a fundamental ecosystem service provided by these landscapes.

Additionally, it is important to recognize that some rangelands occur in wetter, cooler regions or have soil and geologic characteristics that enable them to provide or store additional water. Surface and groundwater provide crucial benefits to habitat for aquatic organisms and drinking water for humans, wildlife, and livestock. Wetlands and riparian areas are also rangelands, although they represent a small percentage of the land area. These are the highly biologically diverse areas found close to reliable supplies of water (streams, rivers, ponds, lakes, depressions, etc.). The overarching ecosystem service values of these wetlands are linked to the watersheds in which they occur, and their importance is disproportionate to the land area they occupy. They provide key habitats for numerous plants and animals and support many other ecosystem services (Wilcox et al. 2017; Wohl et al. 2012).

Forested watersheds, alternatively, provide almost two-thirds of the fresh water in the United States, and healthy, well-managed forests provide natural filtration of precipitation and provide many downstream benefits, lessening the negative impacts from stormwater runoff. As such, forest restoration projects near population centers are increasingly being viewed as an opportunity to protect municipalities from effects of extreme weather events.

Opportunities for land stewards

Voluntary, incentive-based programs can be a valuable option for landowners interested in protecting and restoring rangelands and forests to improve water quality and quantity. For example, there are opportunities within the US Farm Bill prioritized to protect native rangeland and convert croplands and/or production timberlands with resource-conserving vegetation covers in exchange for yearly payments (USDA 2023). These programs prioritize improving water quality and reducing erosion. Other opportunities include private conservation efforts, such as the Ducks Unlimited Preserve Our Prairies initiative, which aims to protect vital wetland habitat for waterfowl in the Prairie Pothole Region of North America. Largely, water quality and/or quantity opportunities for landowners are very localized, contextual to specific watersheds or regions, and tied to specific water conservation outcomes.

Carbon

Carbon is an essential element of life on our planet and is found in all facets of our lives from the land we stand on, the food we eat, the air we breathe, and the clothes we wear. However, when most people think of carbon on the planet, they think about carbon in the form of CO_2 in the atmosphere. Although atmospheric carbon concentrations are important to many biogeochemical processes, the atmosphere is not the largest carbon pool on the planet. For example, the earth's atmosphere contains approximately 750 petagrams of carbon (Pg C), total terrestrial vegetation contains 580 Pg C, and the top three feet of soil holds approximately 1,417 Pg C (not including soil inorganic carbon); thus, just the top three feet of our soil contains more carbon than the vegetation and atmosphere combined (FAO 2017).

Soil carbon is composed of both inorganic (SIC) and organic (SOC) fractions. Soil inorganic carbon is derived from weathering parent materials and carbonates, while soil organic carbon is derived from decomposing tissues from dead plants and animals and soil organisms. With rangelands comprising ~40% of the earth's terrestrial surface and storing 10%–30% of global soil organic carbon (Conant 2012), rangelands play a key role in global carbon cycling.

Soil carbon sequestration is the process by which carbon is fixed from the atmosphere through plants and other microbial processes, then stored in the soil profile. The amount of soil carbon sequestration possible at a location is dependent on soil texture, the climate (temperature and precipitation regimes), the type of vegetation present (perennial vs. annual and woody plants vs. grasses); however, the management being applied and weather can change short-term fluxes (Sanderson et al. 2020). Because these variables (among others) can impact soil carbon sequestration potential, working lands that cover a vast area of the earth's terrestrial surface have considerable variability in their potential to sequester carbon. Rangelands that receive more rainfall (>20 inches of annual precipitation) are poised to sequester more soil carbon annually than drier rangelands (<20 inches of annual precipitation) (De Deyn et al. 2008). These differences in potential do not mean drier rangelands are less important regarding soil carbon sequestration but actually the opposite: Xeric rangelands are vast and can maintain and secure soil carbon if managed to maintain healthy soils and plant communities (FAO 2017). Similarly, according to the USDA Northern Research Station, forestland, harvested wood products, woodlands, and urban trees collectively represent the largest net carbon (C) sink in the United States, offsetting more than 11% of total greenhouse gas (GHG) emissions annually (Domke et al. 2020). Sustainably managed forests that promote the use of prescribed fire, selective harvest, and regeneration, in addition to the tree benefits, generate similar benefits to rangelands due to their diverse native understory plants.

Thus, the sequestration of carbon has been recognized as a tool to mitigate the effects of a changing climate. Over 1.2 billion tons of carbon could be stored annually in working land soils (cropland, grassland, and forests), representing a significant sink that could potentially compensate for a portion of global greenhouse gas emissions (Lal 2016; Chabbi et al. 2017; Dignac et al. 2017; Minasny et al. 2017). The global potential of SOC sequestration and restoration of degraded soils is estimated at 0.6 to 1.2 Pg C per year (Pg C/y) for about 50 years with a cumulative sink capacity of 30 to 60 Pg (Lal 2003). For these reasons, applying management in agricultural production systems to rebuild soil carbon concentrations could provide significant benefits to the production subsystem and society.

Opportunities for land stewards

Increasing the soil's capacity to sequester carbon is a function of several biotic and abiotic factors as outlined previously. However, foundationally increasing the concentration of SOC in soil is a function of increasing the concentration of soil organic matter. Soil organic matter is 58% soil carbon. Soil organic matter is governed by many of the same functions that drive soil carbon accumulation. However, there are several management approaches that have demonstrated the ability to positively impact soil organic matter formation. The following soil health-building principles are agnostic to land use:

1. Understand your ecological context.
2. Keep the ground covered.
3. Minimize unnatural soil disturbance.
4. Promote diversity.
5. Keep living roots in the soil year-round.
6. Practice adaptive livestock management.

Biodiversity

Working lands can vary significantly in the degree to which they harbor plant and insect diversity. For instance, irrigated pasture systems and integrated cropland-livestock systems and production forestry are intensively managed and tend to support lower levels of plant diversity than extensive rangeland systems do. Unlike intensively managed pastures and croplands that largely focus on monoculture crops, actively managed native working lands often include relatively diverse plant communities and a high number of native plant species. Plants provide sources of food, shelter, cover, and services for animals that humans care about, including pollinating insects, birds, and game species. Insects often rely on specific plants that are active at a certain time of year. As a result, more-diverse plant communities tend to support more-diverse insect communities.

Insect diversity can support pollination, the process by which pollinator insects transfer pollen grains from the male anther of a flower to the female stigma of another flower of the same species, which leads to seed production. While wind facilitates this task for many plants, for many others, insects typically serve as the facilitator that moves pollen from one flower to another. Fortunately for the insects, as they collect pollen, they consume some for its protein value or feed on a flower's nectar for energy.

Native pollinators include over 3,500 species of bees, butterflies, moths, wasps, flies, and beetles (and bats and birds). Native pollinators live, forage, and breed in the wild environments and rarely become domesticated. Some native pollinators depend on a single or a limited number of plant species for their needs. As a result of these specializations, there are often direct relationships between plant and native pollinator biodiversity.

Working lands, such as diverse native rangeland ecosystems, create symbiotic relationships with pollinators, and this symbiosis presents a challenge for distinguishing the benefits of plant and pollinator diversity. The presence of pollinators on the landscape is a function of the availability and diversity of floral resources. Ultimately these mutualistic relationships are dependent on each other, meaning if management practices that limit plant diversity are implemented (e.g., herbicide application), the resulting unintended consequences may be reflected on the pollinator community as well.

Insect pollination plays a critical yet unassuming role in the success of the agricultural industry. Approximately 35% of the world's food crops either require or benefit from insect pollination (USDA NRCS 2016). Ultimately the health and function of working lands depend on complex relationships between plants, soils, animals, and increasingly the impacts and services provided by pollinators.

Opportunities for land stewards

Opportunities are currently limited for land managers to receive payments from society when taking actions to support biodiversity services. However, these markets are emerging. There are nascent markets as of 2023 in New Zealand, Colombia, and Australia. Many of these markets are localized as they are focused on increasing biodiversity of specific habitat types intended to benefit imperiled wildlife species. However, some are being created with a stacking feature to be coupled with a soil carbon credit. Stacking refers to the opportunity for various overlapping ecosystem services produced on a given piece of land that are measured and separately "packaged" into a range of different credit types or units of trade that together form a stack or bundle (Von Hase and Cassin 2018). Voluntary markets for biodiversity credit generation in the United States are largely yet to be developed. Although no ecosystem service credit framework for biodiversity currently exists, there are financial assistance opportunities for land managers interested in managing working lands for increased biodiversity.

Focused on conserving cropland, USDA's Conservation Reserve Program (CRP) provides financial incentives for producers interested in conserving highly erodible cropland by planting diverse native species mixtures, sometimes with the goal of creating habitat for pollinators. Commercial seed companies provide additional opportunities for landowners through the collection and/or increase and sale of seed from species used in native plant restoration.

Opportunities also exist to create enterprise diversification of existing operations. For instance, working lands located in areas near croplands may be able to capitalize on plant diversity present in existing rangelands by adding a honeybee business to their operation. A significant proportion of honeybee colonies in the United States are moved to the Northern Plains over the summer to forage on grassland flowers in preparation for overwintering to provide pollination services and to produce honey, valued at $343 million (USDA NASS 2017). There are also emerging opportunities for producers to market products with pollinator-friendly certifications.

Wildlife Habitat

All living species need unique combinations of food, cover, water, and space. The combination and distribution of these components ultimately define the usefulness of an area for wildlife species. Individual wildlife species require specific combinations of food, cover, water, and space to define their habitat needs, although some species are more opportunistic than others. The complexity of working lands provides the foundation for these terrestrial habitats. Working lands provide biotic resources, such as plant diversity for wildlife habitats, through inherently variable soils and resultant ecological sites. Working lands support a greater number of plant and animal species because they comprise unique topographic features, soils, and precipitation regimes that add to structural complexity, providing additional spatial and temporal niches. Therefore, working lands managed adaptively for multiple outcomes associated with the provision of wildlife habitat (e.g., vegetation height and structure) provide a diverse suite of habitats that benefit multiple wildlife species and in return produce a suite of wildlife-related ecosystem service benefits.

Many land managers and the general public benefit from working lands because they support wildlife species of importance that provide consumptive (hunting and fishing) and nonconsumptive (viewing) values from recreational opportunities. The 2016 National Survey of Fishing, Hunting, and Wildlife-Associated Recreation showcased that 103.7 million Americans participated in wildlife-related recreational activities. These recreationalists spent $157 billion with $76 billion on wildlife watching, $46 billion on fishing, and $26 billion on hunting. At a national scale, the status and condition of both consumptive and nonconsumptive wildlife populations serve as a general indicator of the overall health and condition of the land. Essentially, when wildlife populations thrive, the value of the land, in terms of ecosystem services, typically thrives too, as there are synergistic interactions between the two. However, some wildlife species and their related habitat are in peril because of land-use changes and pressures. Simply identifying ecosystem services fails to create or incentivize the production of these services.

Opportunities for land stewards

There is growing interest in management approaches that provide mutualistic benefits, for instance, strategies that optimize productivity and profitability of the agricultural production system while simultaneously fostering wildlife habitat. Practices such as patch burning and adaptive grazing management with specific objectives for structured vegetation heights benefit many wildlife species, especially grassland birds. Marketing opportunities are emerging for producers interested in prioritizing grassland bird conservation, such as Audubon's Conservation Ranching Initiative that encourages customers to make a difference by purchasing their beef from grassland bird–friendly lands. Ultimately, the market driver for the creation and management of wildlife habitat is the ability for landowners to re-create on their own land or sell that right to access as a hunting lease. In Texas, bobwhite quail leases can exceed $20 to $25/acre. Hunting lease values for white-tailed deer are similar. As stated earlier, nonconsumptive uses are a $76 billion industry in the United States, which is growing fast. These provide considerable diversification investment opportunities for many beef cattle operations, although this diversification and the subsequent financial gain comes with the additional requirement to focus management strategies that optimize both enterprises rather than maximize one over the other.

Challenges Facing Ecosystem Service Markets

Today, many questions remain about the implementation of ecosystem service markets (both offset and inset). For any financial mechanism to work, landowner adoption is paramount. Early adopters will set the stage for perceptions about ecosystem service markets. If early adopters have poor experiences, it will hinder the cycle of adoption among landowner peers.

A limited number of experts are exploring the legalities of ecosystem contracts and the corresponding challenges. A pioneer in defining legal contracts for ecosystem services is Texas rancher and attorney Anson Howard. In a 2022 Texas Ag Land Trust (TALT) podcast, "Beyond the Fence Line," Howard joined TALT CEO Chad Ellis to discuss a variety of issues impacting producer adoption and implementation of ecosystem service contracts.

Generally, Howard suggests that the science is not keeping pace with the development of ecosystem service markets. The ecosystem service market is largely focused on carbon, and little has been done on additional ecosystem services. With the paucity of science, some brokers pay the landowner to implement certain practices on the land despite the limited scientific evidence indicating whether these practices will store more carbon, while other brokers pay for actual carbon storage. Howard notes that most contracts tend to be 15 years in length. He states that if a landowner breaks a contract, the landowner is typically responsible for carbon credits that were sold, plus the measurement costs. Contracts also vary in whether they require the landowner to pay for the cost associated with measurement and verification. Some require the landowner to cover that cost up front, while others deduct it out of the landowner's payment for the ecosystem services. In many instances, this reduction can consume half of the value of the ecosystem service payment to the landowner. Further, Howard and Ellis noted that depending on the contract, landowners may not be able to obtain payment for additional carbon stored over what was predicted, as some brokers capture this value but do not share it with the landowner. On the other hand, concerns also exist regarding what happens if the projected quantity of soil carbon is not stored. Another concern highlighted was the ability of producers to participate in pasture and crop insurance and in other ecosystem service stacking benefits and government programs while under some contracts. Howard goes on to highlight that there is no shortage of questions and concerns from producers as they consider participating in ecosystem service markets (Texas Ag Land Trust 2022).

The Future of Ecosystem Service Markets

In the last decade, ecosystem service markets have developed rapidly compared to previous work in the area. Many challenges have begun to be addressed, but there is still a need for a great deal of work to be done in this field to bring clarity. As different administrations take office, the focus and pressure in this area may change, and the focus has certainly increased dramatically under the Biden administration. It is unlikely, though, that the pressure from younger generations will decline. Gen Z and beyond will continue to apply pressure on companies and ultimately not only make purchasing but also investment decisions based on how companies address sustainability and ESG. As this space continues to evolve, it may eventually shift from a predominantly offset to inset market, or from voluntary to a regulated space as well, which could substantially impact the market price for ecosystem service credits. Looking into the future, innovations could be spurred through new incentives and initiatives, like specific lending-based pricing or IRS tax credits for "philanthropic conservation."

Conclusion

Sustainability and ESG are now commonplace considerations at both the consumer and corporate levels. This focus has created a demand for ecosystem services (provisioning, supporting, regulating, and cultural) that private lands can and must help meet. Public lands alone cannot provide all the ecosystem services that society needs. Private lands not only provide food, fuel, and fiber, but they are critical to ensuring clean and abundant water, carbon sequestration and storage, and biodiversity of plant and animal life on earth. Management of these private lands is critical in increasing and delivering these ecosystem services. Further, these private lands provide countless economic benefits to the surrounding communities. Yet private landowners face financial challenges in being able to manage their lands in a way that places more focus on ecosystem services. Financial mechanisms continue to evolve to help address this, but they are far from perfect. Creative and innovative thought leaders are needed in this area to help reward the private landowner who conserves the public interest.

SECTION II

Trends and Status of Private Lands

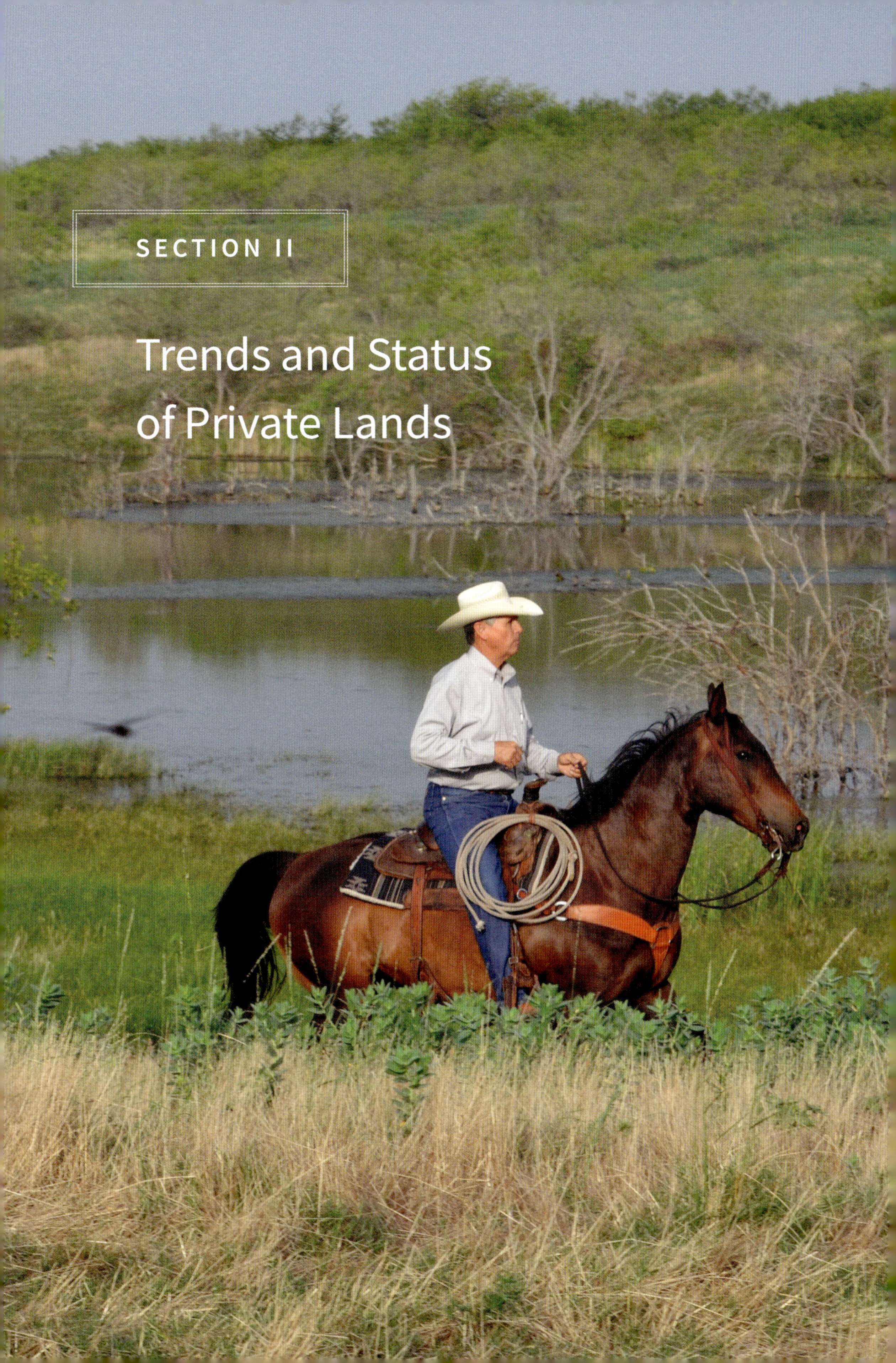

CHAPTER 4

Land in the United States

National Overview

ADDIE SMITH, MATTHEW CRAWFORD, KEVIN SKOW,
GARRETT POWERS, ALISON LUND, AND ROEL R. LOPEZ

The geographic scale and diversity of private lands in the United States are vast. This chapter presents a national overview of the major factors differentiating private lands and reviews trends in land use, land cover, ownership status, productivity metrics, and other drivers for economic and cultural differences in how private lands are managed across the country.

Introduction

Privately owned working lands provide substantial economic, ecological, and recreational resources across the United States. Nearly three-fourths of the entire United States is under private ownership. For these reasons, sustaining the value and public benefits derived from private lands needs to include engaging the land steward and understanding the makeup and characteristics of the land itself. This section is divided into two primary components. First, we provide the reader with a broad overview on the current status of land ownership in the United States, including general land uses, public versus private lands, and the identification of private working lands, which is the primary focus of this book. Due to the unique nature of private working lands, we illustrate differing attributes and characterizations through the division of the continental United States into seven regions (figure 4.1): Pacific Northwest, Southwest, Northern Plains, Southern Plains, Midwest, Southeast, and Northeast. Each region is grouped by states similar in location, climate, ecological characteristics, and agricultural industries. By disseminating private working land data together as a whole, and also separately within regions, we hope to illustrate and characterize the benefits of private lands in the United States to the general public, as well as some of the likely threats to their management in future years.

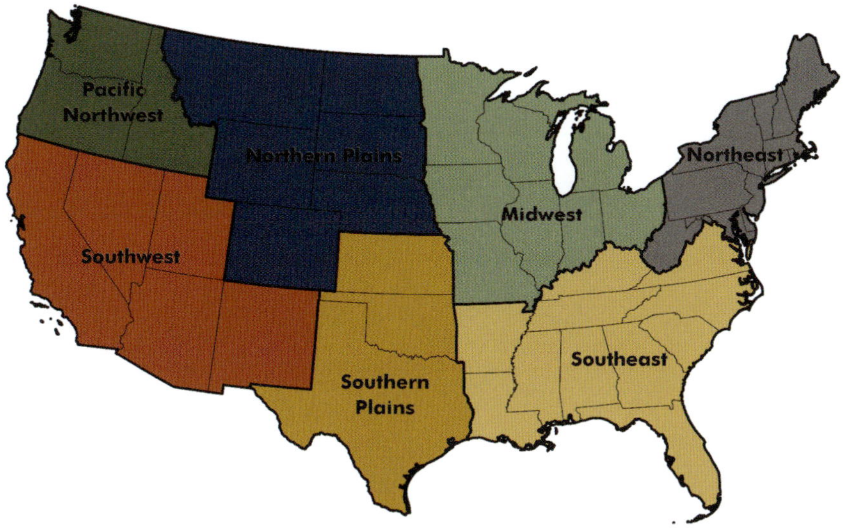

FIGURE 4.1. US regions used in characterizing private lands.

Land Use

The conterminous United States includes over 941 million acres, or about 1.47 million square miles, of a multitude of rural landscapes and land uses that ultimately translate to varying agricultural enterprises and economies (Conservation Science Partners and American Farmland Trust 2020b). As climate and ecological characteristics shift across the country, so do the industries and land uses that can be sustained from them. We begin our effort to describe land ownership nationally by describing land cover and land uses focusing on agricultural land (figure 4.2). Here we include four distinct agricultural land classes: cropland, pastureland, rangeland, and woodland associated with farms in our description. We also included federal lands used for grazing due to the important role that it plays primarily in western states. Our mapping effort contains a new land-use class, low-density residential, in which the average housing density is above the level where agriculture is typically viable. From our analysis, it becomes apparent that the majority of our nation's rangeland and cropland is found in the Northern and Southern Plains, the Midwest, and along the Mississippi Delta. Forestland dominates the Southeast and Northeast, and much of the

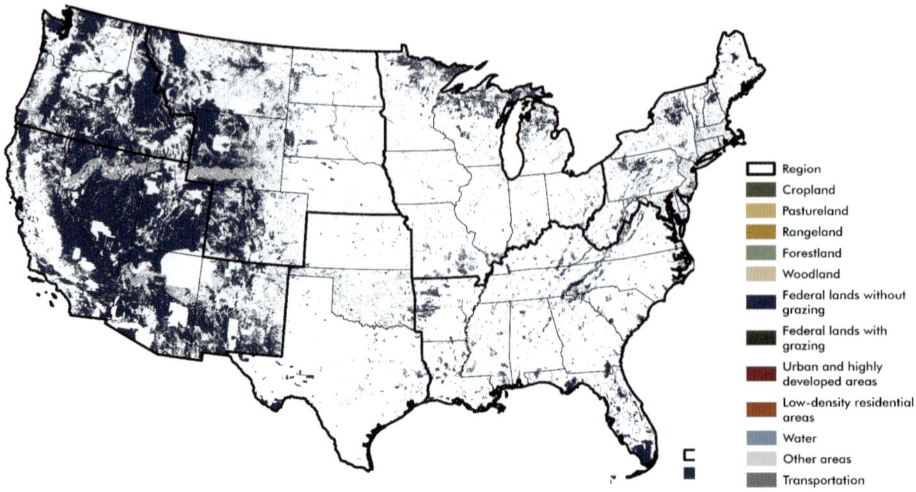

FIGURE 4.2. Land cover and land use by state and region across the United States (adapted from Conservation Science Partners and American Farmland Trust 2020b).

Southwest and Pacific Northwest is federally owned land. This tapestry of both land cover and land use serves to characterize the diversity and breadth of natural resources in the United States.

Public Land

As mentioned previously, one particular land use important to the conservation of working lands is publicly owned protected areas (figure 4.3), including federal, state, and locally owned lands, and also voluntarily provided private protected areas. Approximately 26% of the total landmass for the United States is public lands (US Geological Survey 2018). Starting in 1946 under the Truman administration, the US Department of the Interior's Bureau of Land Management (formerly the General Land Office and the US Grazing Service) began administering land on the basis of multiple use (grazing, timber, recreation, etc.) and sustained yield, while also conserving natural, historical, and cultural resources. Early on, settlers were encouraged to head west with the lure of multiple homesteading acts, ensuring the transfer of public land to those who resided on and cultivated land after a period of time. Eventually, these homesteading acts were repealed and a new policy of retaining public lands in federal ownership was established with the Federal Land Policy and Management Act in 1976 (US Department of the Interior 2020). Thus, the majority of public lands in the United States are held in the western half of the country, specifically

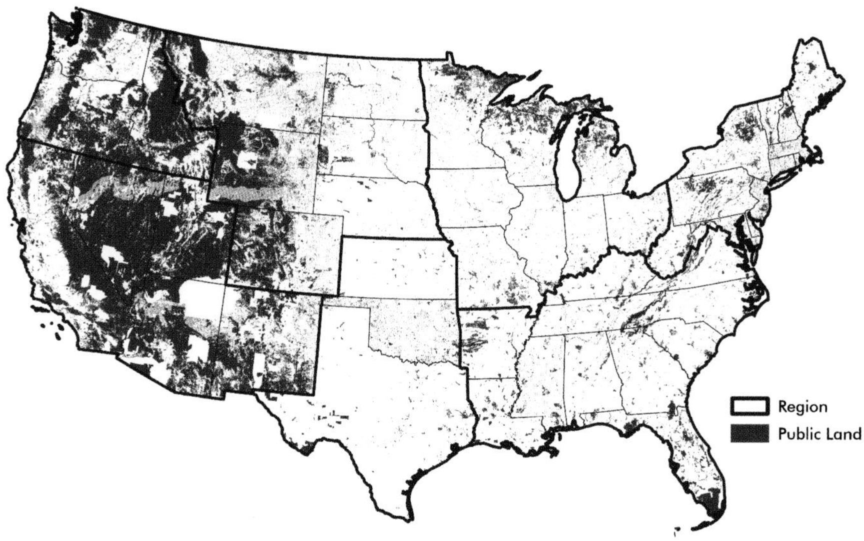

FIGURE 4.3. Public land by state and region in the United States, including federal, state, and locally owned lands as well as voluntarily provided private protected areas (adapted from US Geological Survey 2018).

the Pacific Northwest and Southwest regions, where homesteading had less time to fully take hold. This difference in how land ownership evolved between the eastern and the western United States plays an important role in describing the culture of local and regional communities, to include how private individuals interact with state and federal organizations. This distinction also plays an important role in the view of private property rights between various regions across the United States.

Private Working Lands

While public lands are more prominent in the West, private working lands dominate the Northern and Southern Plains and the Midwest (figure 4.4). Observing private working lands, shown in green, and public lands (both working and nonworking), shown in blue, as a percentage of total state area helps paint the picture of land ownership across the United States. Note these two data sets are independent of each other, so state percentages cannot be combined to depict the state as a whole. Other land ownership types, such as privately owned nonworking lands, are not included. These maps highlight the opportunity and necessity to use different land management and conservation strategies in each state. While publicly

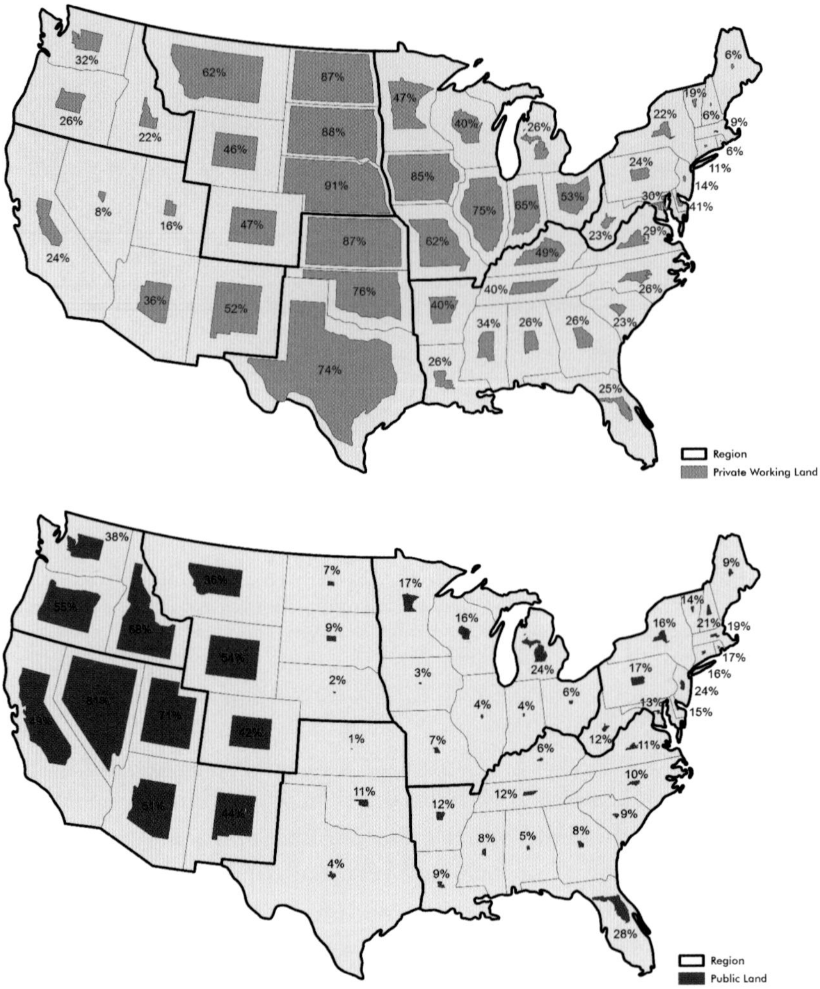

FIGURE 4.4. Percentage of private working land to total state area in the United States (*above, green*) and percentage of public land to total state area in the United States (*below, blue*) (adapted from US Geological Survey 2018; and US Department of Agriculture 2020b).

owned land is often mandated to uphold multiple use and sustainability requirements, privately owned land may employ a different strategy driven by the objectives and goals of the private landowner. Furthermore, public land is generally protected from land conversion to urban uses, which oftentimes threatens private lands.

As part of the "Farms Under Threat" project, the American Farmland Trust created the productivity, versatility, and resiliency (PVR) index data set

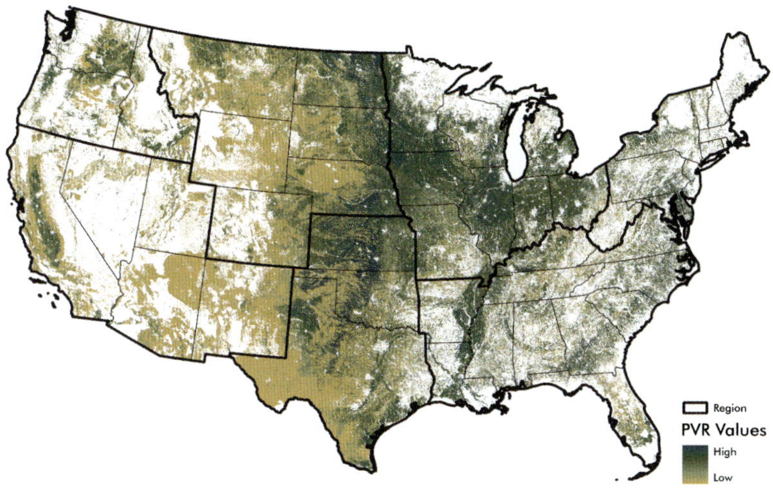

FIGURE 4.5. Productivity, versatility, and resiliency (PVR) index on agricultural land by state and region in the United States. Higher PVR values (*blue*) indicate higher suitability for long-term intensive crop production (adapted from Conservation Science Partners and American Farmland Trust 2020d).

on agricultural land (figure 4.5). The PVR index is the first-ever assessment of agricultural land quality that combines soil productivity and capacity, land cover and use, food production for direct human consumption, production limitations, and length of growing season. Higher PVR values (shown in blue), found primarily in the Northern and Southern Plains and Midwest regions, indicate higher suitability for long-term, intensive crop production, especially for food crops such as fruit, nuts, vegetables, and staples grains (Conservation Science Partners and American Farmland Trust 2020d).

The import ance of this effort is highlighted in observing areas where nonfederal farmland and rangeland were converted to urban and highly developed (UHD) and low-density residential (LDR) land uses from 2001 to 2016 (figure 4.6). According to the American Farmland Trust, from 2001 to 2016, over 10.9 million acres of agricultural land were developed or compromised, seen in red. Of those 10+ million acres, over 6.7 million acres were converted to LDR land use, where scattered, large-lot development fragments the agricultural land base and limits production, marketing, and management options for the working farms and ranches that remain. Farmland (including cropland, pastureland, and woodland associated with farms) with PVR values above the state median are shown in green, and

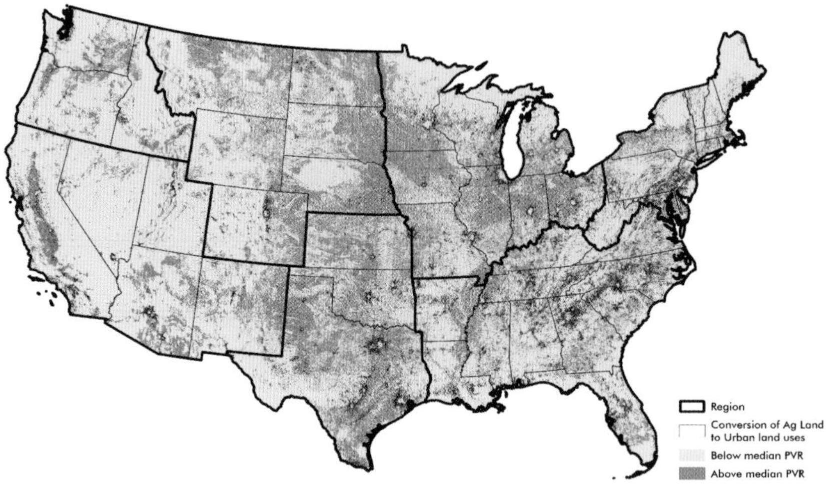

FIGURE 4.6. Agricultural land conversion (*red*) by state and region in the United States from 2001 to 2016, with *above* (*green*) and *below* (*gray*) median PVR index values (adapted from Conservation Science Partners and American Farmland Trust 2020a, 2020e).

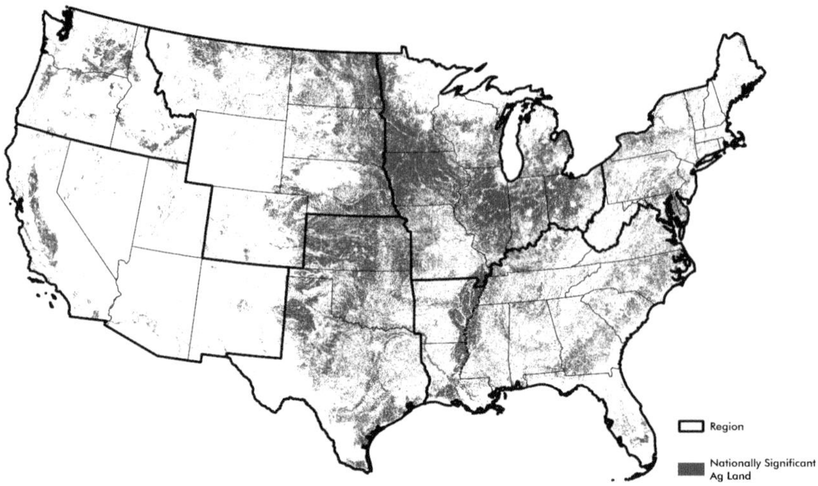

FIGURE 4.7. Nationally significant agricultural land by state and region in the United States (adapted from Conservation Science Partners and American Farmland Trust 2020c).

lands with PVR values below the median are shown in gray (Conservation Science Partners and American Farmland Trust 2020e).

Nationally Significant agricultural land is the land best suited for long-term production of food and other crops (figure 4.7). To identify Nationally Significant land, the American Farmland Trust calculated a minimum PVR value threshold. All agricultural lands with PVR values above this threshold were classified as Nationally Significant, shown in green, found primarily across the Northern and Southern Plains and the Midwest (Conservation Science Partners and American Farmland Trust 2020c). This map highlights the critical areas where conservation efforts should focus.

Together, these maps create a visual depiction of the valuable private lands across the United States and illustrate the threats of loss and conversion that they face. Identifying these areas and their threats is an important step in combating land fragmentation and conversion, and ultimately in conserving valuable working lands for future generations.

CHAPTER 5

Land Status and Trends
Regional Overview

ADDIE SMITH, MATTHEW CRAWFORD, KEVIN SKOW, GARRETT POWERS, ALISON LUND, AND ROEL R. LOPEZ

This chapter examines more specific regional trends in private land ownership and delves into drivers of changes in land ownership and enterprise focus and trends in working land fragmentation and conversion.

Introduction

In the previous chapter, we described some of the broad characteristics of land cover and land use in the United States. From that review, several regional patterns begin to emerge related to primary types of land uses, proportion of private lands to public lands, and areas of agricultural significance, to name a few. Here, we hope to provide a deeper assessment of private working lands by the previously mentioned US regions, with several data sets that help characterize both the current status and trends over time. One of these data sets is collected and maintained by the US Department of Agriculture (USDA). Every five years, the USDA National Agricultural Statistics Service (NASS; 2020a) produces the Census of Agriculture, a complete data set of farms and ranches in the United States and the people who operate them. The Census of Agriculture's ownership data set reports various metrics or characteristics of private working lands. One such measure is the number of farms and acres of farms by size class for each census year (1997, 2002, 2007, 2012, and 2017), which together tell a story of private working lands in the United States (figures 5.1 and 5.2). Regionally, total acres and total number of farms vary greatly. In general, the Northern and Southern Plains and the Midwest contain the largest total acres of working lands, while the Southern Plains, Midwest, and Southeast contain the largest number of total farms. When combined, these two metrics also shed light on historical and ecological trends across the United States, as seen through average farm size (figure 5.3). A gradient

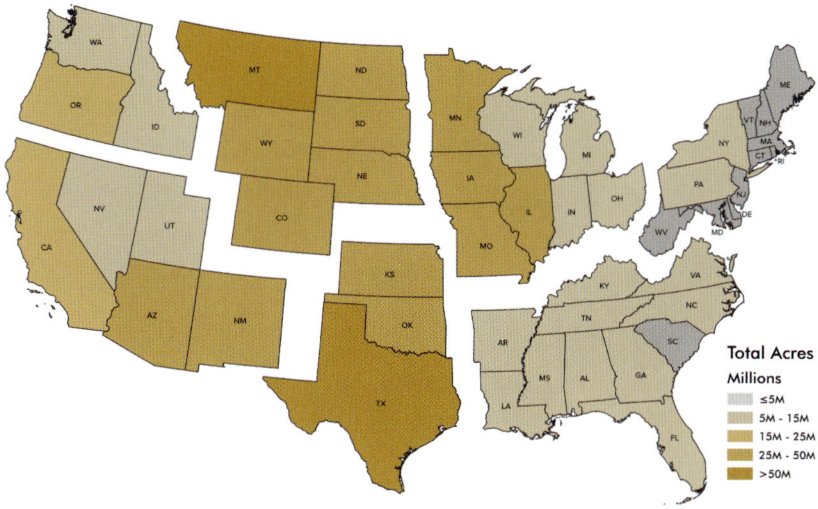

FIGURE 5.1. Total working land acres in the United States in 2017 by state. States are visually separated by regional boundaries.

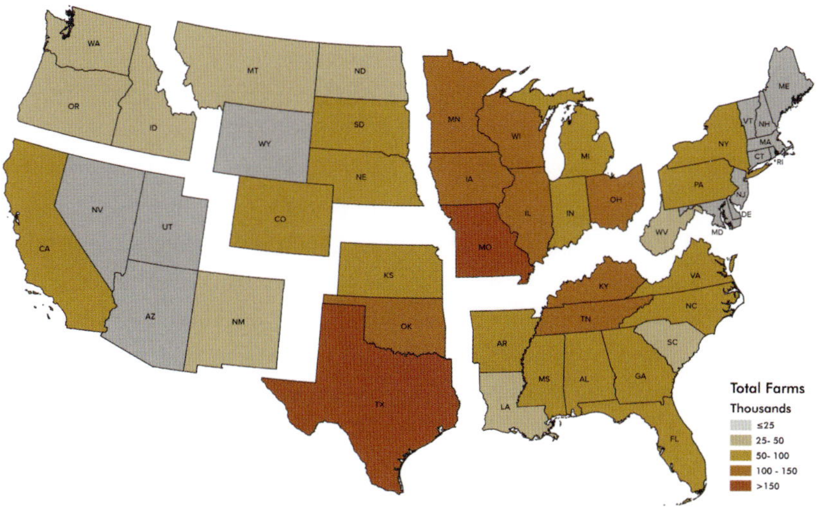

FIGURE 5.2. Total working land farms in the United States in 2017 by state. States are visually separated by regional boundaries.

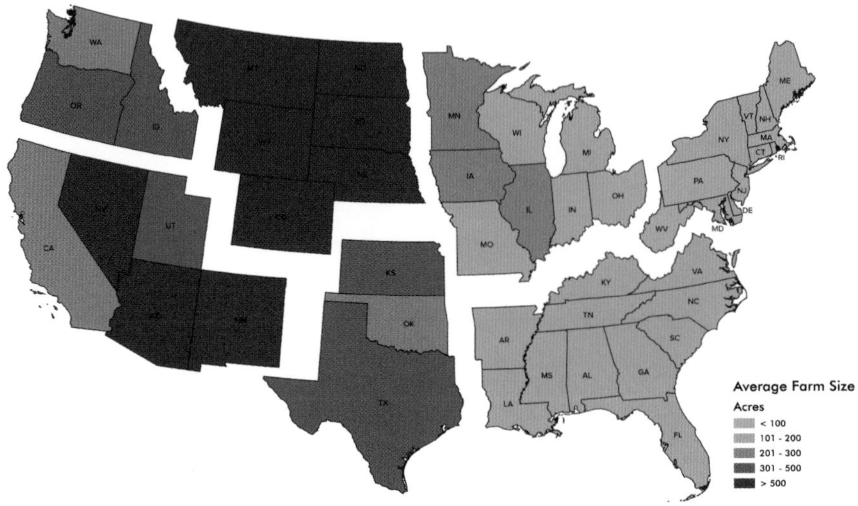

FIGURE 5.3. Average farm size of working lands in the United States in 2017 by state. States are visually separated by regional boundaries.

trend can be seen moving west to east across the United States, with larger farms in the west and smaller farms in the east. The largest average farm sizes are found in the Northern Plains and Southwest regions. This trend is also illustrated when you display average farm size geographically across the regions (figure 5.4). Historically, as the United States grew in size and population, land grants and homesteading laws encouraged western settlement with the promise of land ownership. For ecological and economic reasons, land in the Great Plains and beyond lent itself to larger agricultural operations requiring more land, such as cattle ranching and farming. States and regions with larger average farm sizes often correlate with industries requiring higher acreages of farmland to be profitable (i.e., cattle ranching and farming and oilseed and grain farming).

Agricultural Focus

Another measure that can aid in understanding private land uses is the primary agricultural focus for a given region. The North American Industry Classification System (NAICS) was developed under the direction and guidance of the Office of Management and Budget as the standard for use by federal statistical agencies in classifying business establishments for the collection, tabulation, presentation, and analysis of statistical data describing the US economy (US Census Bureau 2021). NAICS is based on

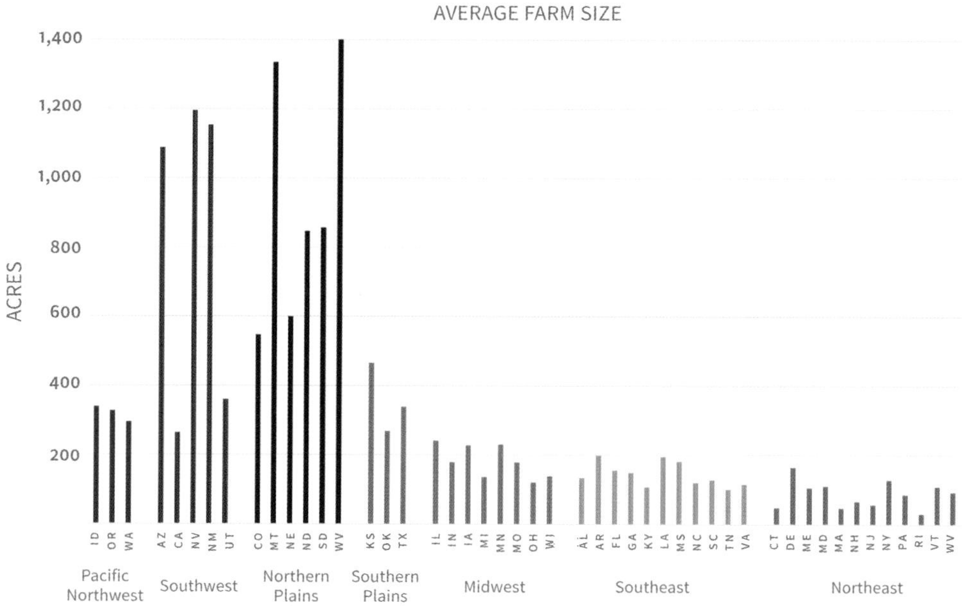

FIGURE 5.4. Average farm size of working lands in the United States in 2017 by state and region.

a production-oriented concept, meaning that it groups establishments into industries according to similarity in the processes used to produce goods or services (US Census Bureau 2021). The NAICS industry maps show the top-ranking agricultural industry by both number of acres (figure 5.5) and number of farms for each state (figure 5.6). The top industries by number of acres in the United States in 2017 included cattle ranching and farming, oilseed and grain farming, other animal production (including apiculture, horses and other equine production, furbearing animal and rabbit production, and all other animal production), and other crop farming (including tobacco, cotton, sugarcane, hay, sugar beet, peanut, and other miscellaneous crop farming). The top industries by number of farms in the United States in 2017 included cattle ranching and farming, fruit and tree nut farming, nursery and floriculture production, oilseed and grain farming, other animal production, and other crop farming. The top agricultural industries vary by region across the United States and often correlate with average farm size. Interestingly, the top industries by number of acres do not always align with the top industries by number of farms. This can be attributed to the differences in land requirements for different agricultural pursuits; for example, a cattle ranch requires more acreage to maintain a profit than does a greenhouse or nursery. It is important to note that these industry classifications focus entirely on traditional agricultural industries and do not include forestry-specific ownerships and their related industry products (Sass et al. 2020). This is an important caveat, as private forest ownership is particularly important to the rural economies and job markets of the Southeast and Northeast regions (figure 5.7; US Department of Agriculture, Forest Service 2020).

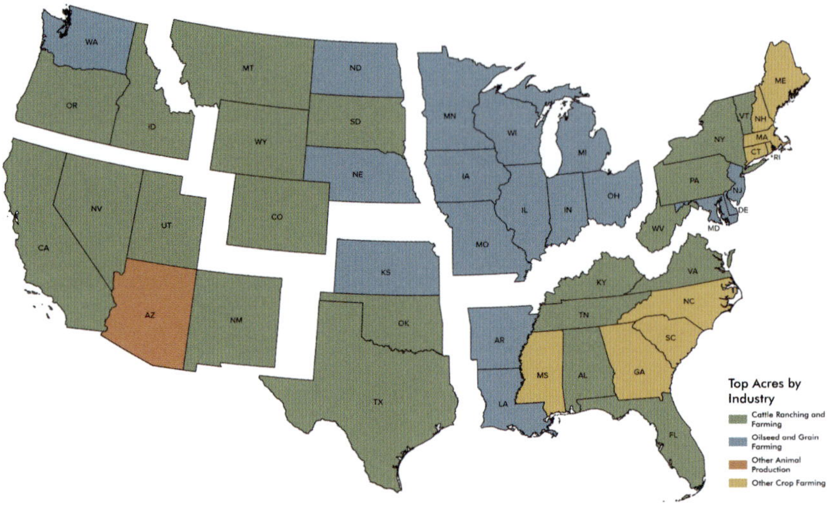

FIGURE 5.5. Top-ranking agricultural industry, using the North American Industry Classification System (NAICS), by number of acres for each state in the United States. States are visually separated by regional boundaries.

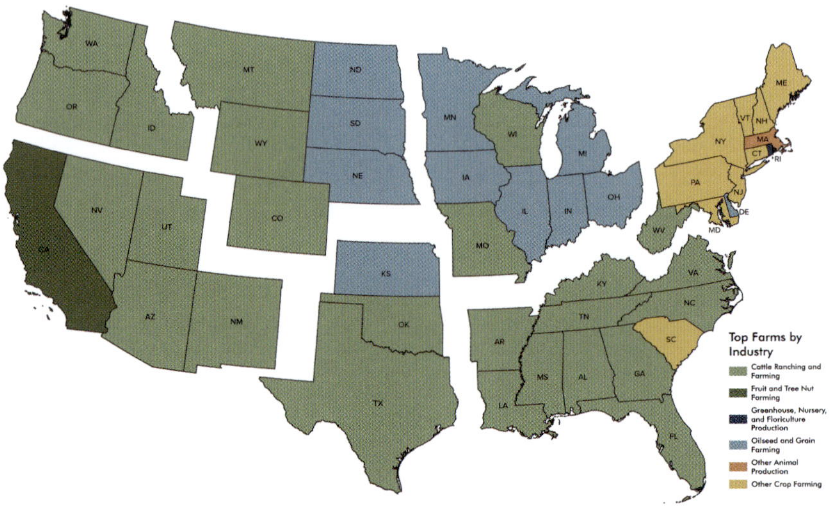

FIGURE 5.6. Top-ranking agricultural industry, using the North American Industry Classification System (NAICS), by number of farms for each state in the United States. States are visually separated by regional boundaries.

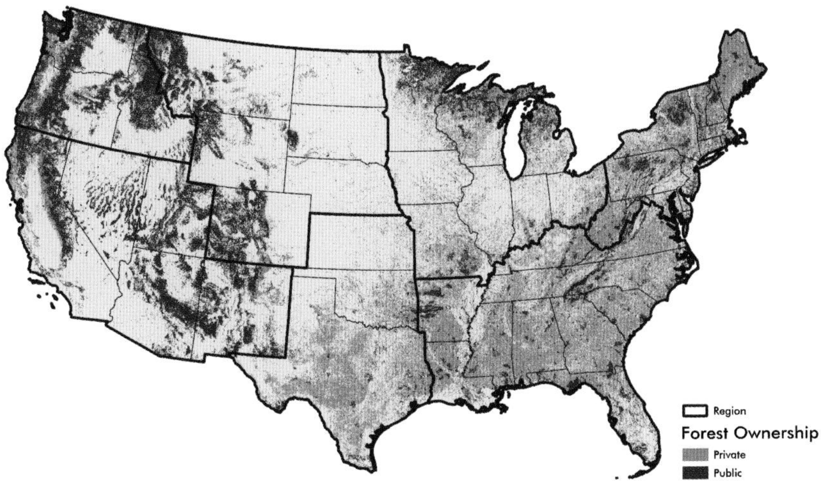

FIGURE 5.7. Private and public forestland ownership by state and region in the United States (adapted from Sass et al. 2020).

Drivers of Change

Thus far, we have described the current makeup and status of private working lands in the United States. These figures and explanations are a snapshot in time and are subject to changes due to increasing pressure for land. Historically, population growth has played a major role in the decline of private working lands across the nation (see figure 4.6). In fact, increasing population growth often is viewed as an early predictor of a loss of private working lands. As urban centers grow in population, the demand for surrounding rural land grows with it. Land asset values have also served as a strong indicator of this demand for rural land. Urban economic growth often leads to population growth, which increases the demand for rural land and drives land values upward. Higher rural land values provide landowners with the incentive to subdivide or sell their property at a greater profit than they can make with traditional agricultural practices. This cyclical process greatly increases the likelihood of ownership fragmentation, land conversion, and the potential loss of working lands altogether. We define *ownership fragmentation* as the breakup of a larger farm or ranch into smaller sections or parcels. Ownership fragmentation can present challenges for the application of certain land management practices, such as prescribed fire. The coordination of a prescribed fire in a landscape of smaller parcels of both working lands and nonagricultural land uses makes this management practice a challenge. *Land conversion* is defined as a change in land use from an agricultural use to a nonagricultural use (e.g., residential or urban).

Private working land asset values vary across the United States but are most notable along the East and West Coasts and the Midwest region (figures 5.8 and 5.9). To compare, some of the more expensive rural asset

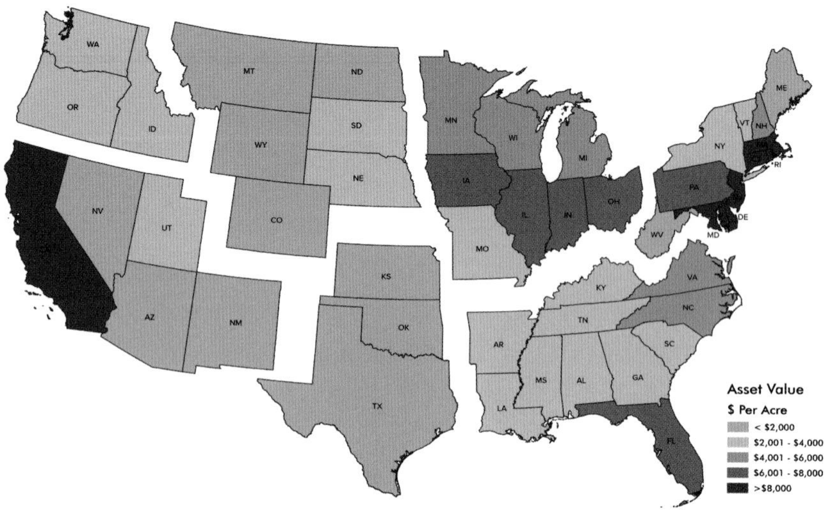

FIGURE 5.8. Working land asset value in $ per acre in the United States in 2017 by state. States are visually separated by regional boundaries.

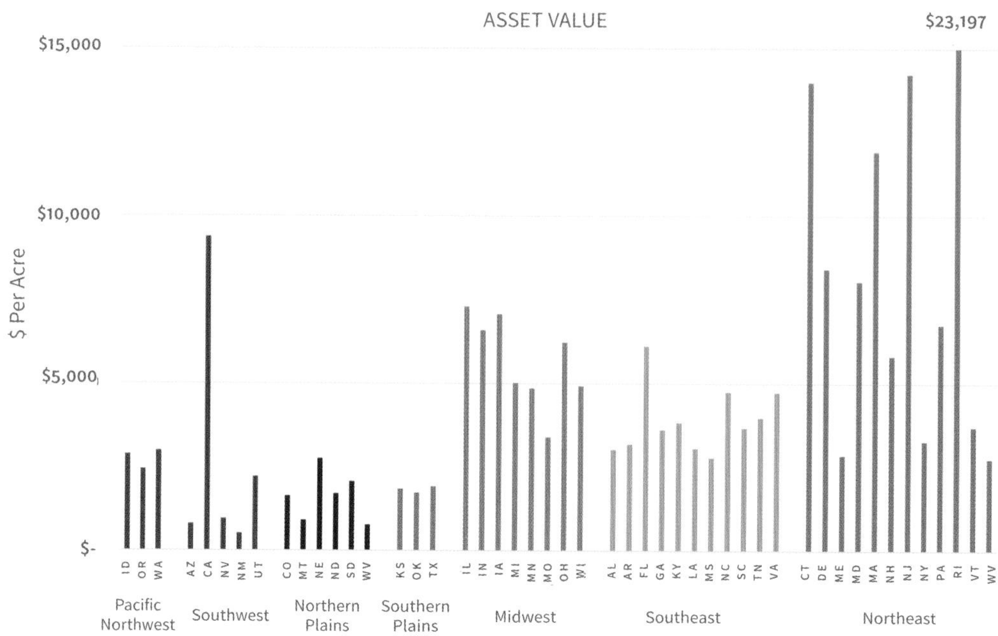

FIGURE 5.9. Total working land asset value in $ per acre for the United States in 2017 by state and region.

value areas in the country (Northeast and Southeast) are also areas that bring in lower farm-related income (figures 5.10 and 5.11). *Farm-related income* includes gross income from farm-related sources received in 2017 before taxes and expenses from the sales of farm by-products, as well as

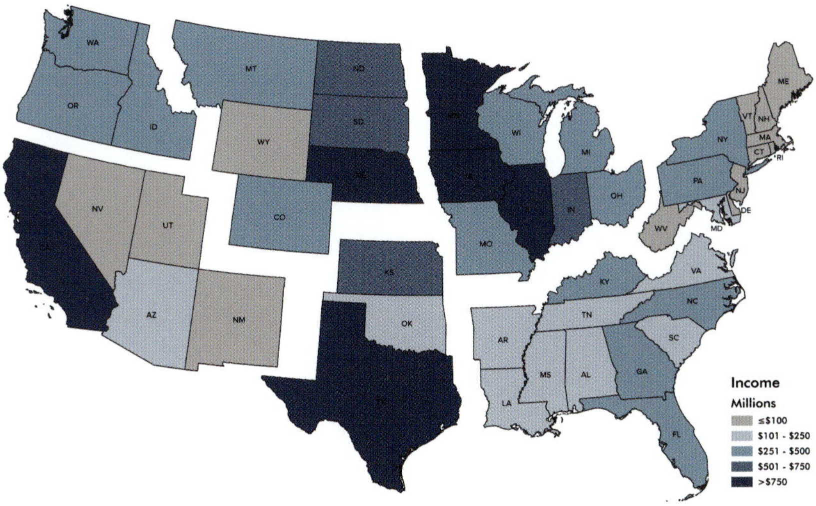

FIGURE 5.10. Total farm-related income in millions ($) for the United States in 2017 by state. States are visually separated by regional boundaries.

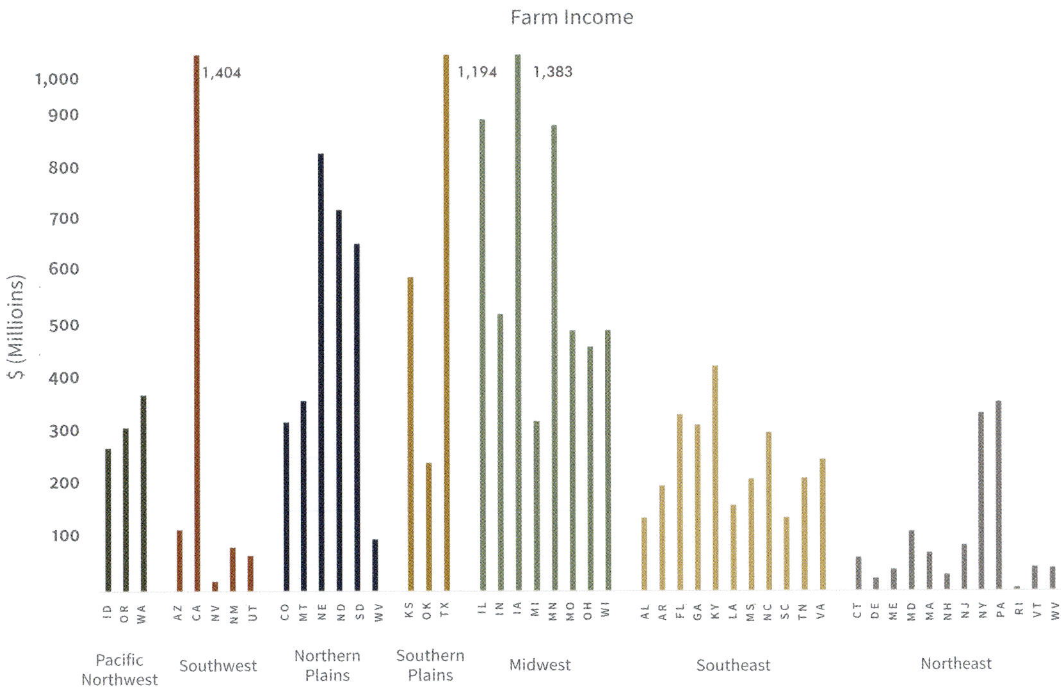

FIGURE 5.11. Total farm-related income in millions ($) for the United States in 2017 by state and region.

other sales and services closely related to the principal functions of the farm business. In addition, smaller average farm sizes in these two regions further exacerbate the conflict of farm profitability. Together, these two factors, relatively static farm-related income values and rapidly rising land asset values, are important underlying drivers of ownership fragmentation and land-use conversion, particularly in the areas within and surrounding growing urban and suburban centers.

Trends in Private Working Lands

Another common measure used to track or observe the loss of private working lands is the percentage change in total acres of working lands (1997–2017; figure 5.12). Across most of the country, private working lands have decreased over the last 20 years. Some regions, such as the Southeast, Midwest, and Northeast, have seen dramatic changes to their total working land base. Additionally, the percentage change in the total number of farms (1997–2017) illustrates trends in the fragmentation of working lands over time (figure 5.13). In every state across the United States, the total number of farms increased, with the most dramatic changes occurring in the Northern and Southern Plains and the Southwest. Combined, these two values paint the picture of the continued loss of private working lands in the United States; collectively more and more farms are being added to a decreasing working land base. Figure 5.14 visualizes this trend—across most states working land acres are decreasing (yellow) and working land farms are

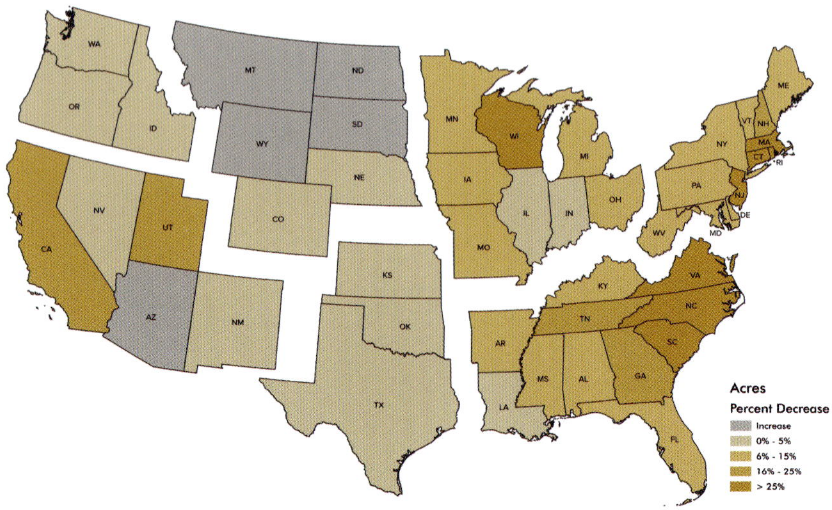

FIGURE 5.12. Loss of working lands seen through the percentage decrease in total acres of working lands from 1997 to 2017 by state in the United States. States are visually separated by regional boundaries.

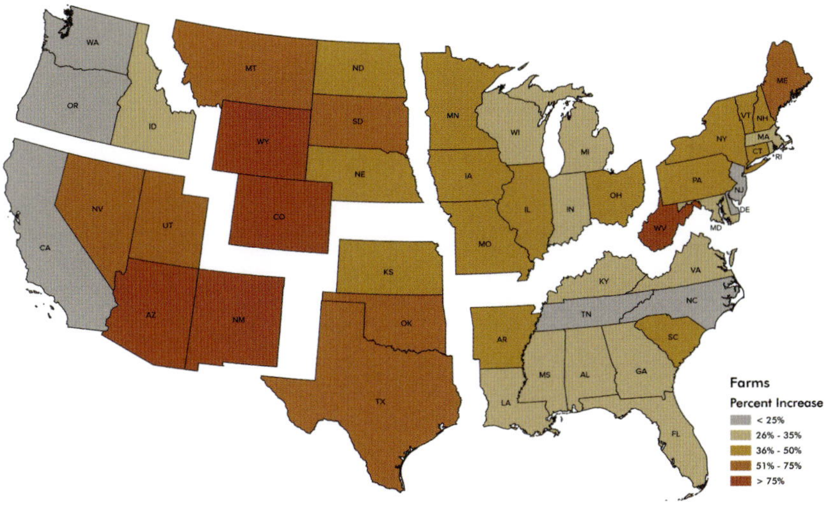

FIGURE 5.13. Fragmentation of working lands seen through the percentage increase in total number of working land farms from 1997 to 2017 by state in the United States. States are visually separated by regional boundaries.

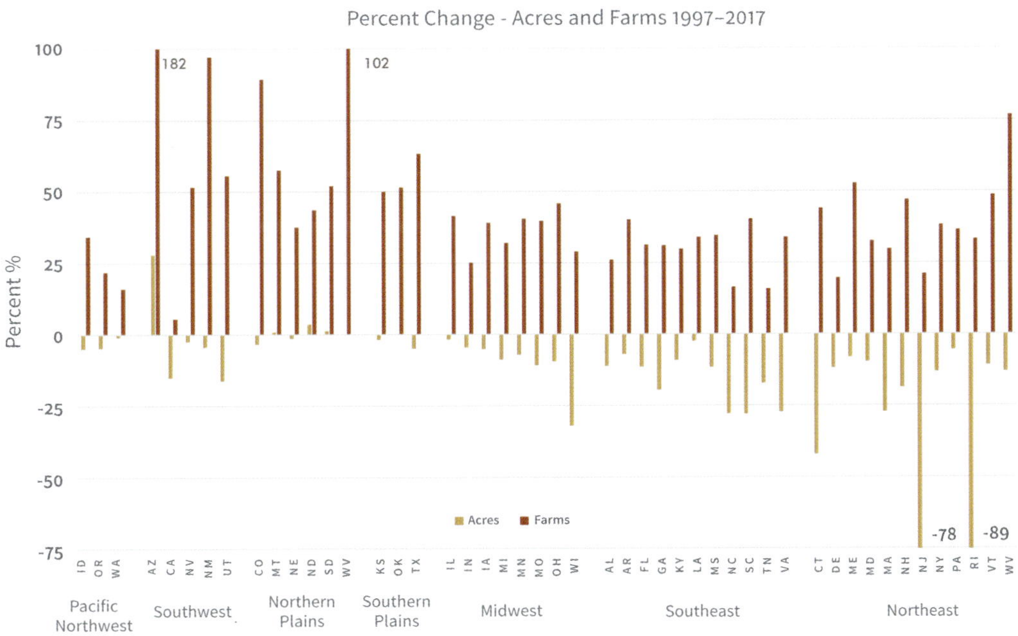

FIGURE 5.14. Percentage change in acres of farms (*yellow*) and number of farms (*orange*) from 1997 to 2017 in the United States by state and region.

increasing (orange). Where drivers of land-use change are present, farms are getting smaller, potentially less profitable, and the functionality and productivity of surrounding working lands are decreasing.

Conclusion

This chapter provides a broad overview on the current status of land ownership in the United States and also identifies key trends in private working lands. Privately owned working lands provide substantial economic, ecological, and recreational resources across the United States that are at risk of fragmentation, conversion, or loss. Effective management and conservation of these lands to sustain public value and benefit will include engaging the land steward and understanding key characteristics of the land itself. By disseminating private working land data across the United States and separately within regions, we hope to adequately tell the story of the benefits of private lands to the general public, to identify some of the likely threats to their management in future years, and to highlight effective approaches to protecting this valuable resource for future generations.

SECTION III

Trends and Status of Private Landowners

CHAPTER 6

General Demographics

ANGELICA LOPEZ, DANIELA BARRIENTOS, TONY MAYORAL, AND ROEL R. LOPEZ

This chapter explores demographic characteristics of private landowners across the United States and various regions. Demographic characteristics relevant to program development are also included in terms of operation size and economic status. Understanding landowner demographic characteristics is relevant to engaging, developing, and improving natural resource and agricultural-related programming.

Introduction

Private landowners (hereafter, collectively referred to as landowners or producers) and managers play a significant role in the stewardship of private lands (family farms, ranches, and forestlands), which provide economic, ecological, and recreational benefits across the United States. Unlike public lands (government or state-owned), each unit of private land or farm is referred to as an *operation* by the United States Department of Agriculture (USDA). Operations are further classified by legal status: family and individual agricultural operations (86%), partnerships (6%), agricultural corporations (6%; of these, 89% are family held), and institutional, research, reservation, and other operation types (2%). About 60% of the United States' landmass is under private ownership, and approximately 3.7 million landowners manage unique operations, bringing different perspectives and experiences to their land stewardship practices. In this chapter, we provide both a national and regional demographic overview and comparison of landowners based on data from the USDA's Census of Agriculture, which links producers to their unique land-based operations. We describe general demographic characteristics, such as sex, age, and race/ethnicity, along with associated demographic descriptors, such as internet service, military service, years of operational experience, size of farm, and economic status. Collectively, these illustrate and characterize rural communities and the potential challenges associated with stewarding private lands. Land stewardship falls on a continuum, and operational decisions made by landowners influence not only their land but also local communities via public or nature benefits derived from the land. Johnston (2018) defines ecosystem (nature) benefits as "outputs, conditions, or processes of natural systems that directly or indirectly benefit humans or enhance social welfare," such as clean air and water. Mismanaged lands offer little in the way of nature benefits, whereas those managed with a strong land ethic operate in a way that yields benefits enjoyed by all.

Because regions across the United States vary ecologically, agricultural production outputs, known as commodities, also vary among and within each region. For example, ranching in Montana differs greatly from ranching in Texas; cattle are produced in both locations, yet the circumstances and operational needs will vary. Since the location and uniqueness of landowners influence operational needs, we divided the nation into seven regions grouped by states similar in location, climate, ecological characteristics, and agricultural industries: Pacific Northwest, Southwest, Northern Plains, Southern Plains, Midwest, Southeast, and Northeast (figure 6.1). By summarizing landowner demographic data nationally and by region, we are able to illustrate similarities, differences, and needs over a large geographic scale. Understanding the needs of landowners will assist in developing programs that better support landowners and land stewardship. Finally, the data used for our assessment were provided by the USDA National Agricultural Statistics Service (NASS) Census of Agriculture (USDA NASS 2021). Many thanks are owed to Virginia Harris, Joseph Hagedorn, Thomas Laidley, and everyone who helped us complete this work.

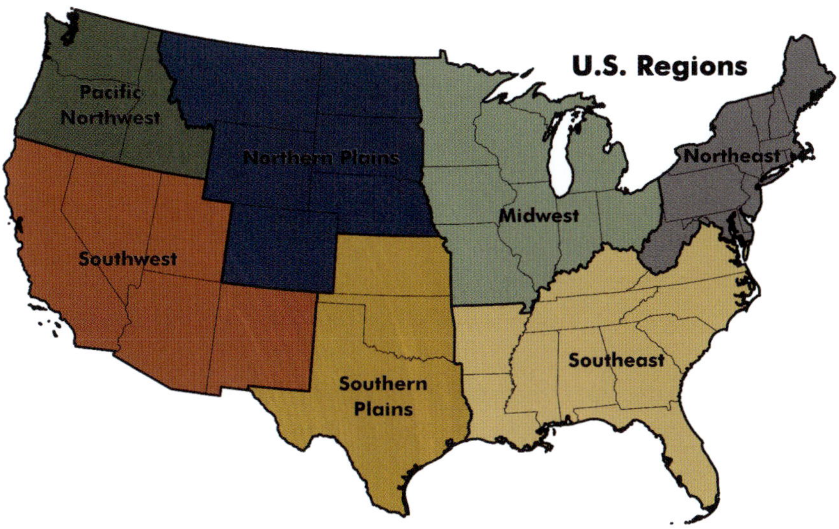

FIGURE 6.1. US regions used in characterizing private landowners.

Note: Terms used in this report reflect data and use by the United States Department of Agriculture, which defines them as follows:

Agricultural Commodity: According to the Code of Federal Regulations, an agricultural commodity is "unprocessed product of farms, ranches, nurseries, and forests and natural and man-made bodies of water, that the Independent Producer has cultivated, raised, or harvested with legal access rights. Agricultural commodities include plant and animal products and their by-products, such as crops, forestry products, hydroponics, nursery stock, aquaculture, meat, on-farm generated manure, and fish and seafood products. Agricultural commodities do not include horses or other animals raised or sold as pets, such as cats, dogs, and ferrets."

According to the Code of Federal Regulations, an agricultural producer (Independent Producer) is:

An individual or entity that produces an Agricultural Commodity through participation in the day-to-day labor, management, and field operations; or that has the legal right to harvest an Agricultural Commodity that is the subject of the Value-Added Producer Grant project.

The Agency shall determine the Agricultural producer status of Tribes and Tribal entities without regard to day-to-day labor, management, and field operation and right to harvest status.

Economic Class of Farms: Economic class data are the classification of farms by the sum of market value of agricultural products sold and Federal farm program payments.

Farms by Legal Status: All farms were classified by legal status in the 2017 census. This section collects information for federal tax purposes to determine an operation's legal status. The classifications used were:

Family or individual (sole proprietorship), excluding partnership and corporation.

Partnership, including family partnership, was further subclassified into:

Registered under State law.

Not registered under State law.

Corporation, including family corporations, was further subclassified into:

Family held or other than family held.

More than 10 stockholders.

Other, estate or trust, prison farm, grazing association, American Indian reservation, etc.

Farm by Size: All farms were classified into size groups according to the total land area in the farm. The land area of a farm is an operating unit concept and includes land owned and operated as well as land rented from others. Land rented to or assigned to a tenant was considered part of the tenant's farm and not part of the owner's.

Farms by Tenure of Producer: All farms were classified by tenure of producers. The classifications used were:

Full owners operated only land they owned.

Part owners operated land they owned and also land they rented from others.

Tenants operated only land they rented from others or worked on shares for others.

Farms with hired managers are classified according to the land ownership characteristics reported. For example, a corporation owns all the land used on the farm and hires a manager to run the farm. The hired manager is considered the farm producer, and the farm is classified with a tenure type of "full owner" even though the hired manager owns none of the land he/she operates.

Hispanic, Latino, or Spanish Origin: Producers of Hispanic, Latino, or Spanish origin are found in all of the racial groups listed in the census and were tabulated according to the race reported, as well as on tables pertaining only to this group.

Institutional, Research, Experimental, and American Indian Reservation Farms: Data for these farms are combined into a single category. Research farms include farms operated by private companies as well as those operated by universities, colleges, and government organizations for the purpose of expanding agricultural knowledge.

Landlord's Share of the Total Sales (Regarding Tenants and Tenancy): Data represent the share of the operation's total sales that went to landlord(s).

New and Beginning Producers: Includes producers operating on any operation for 10 years or less. They may be on farms with producers who are not beginning producers.

Operation: Formally known as a "farm," these are "defined as any place from which $1,000 or more of agricultural products were produced and sold, or normally would have been sold, during the year." According to the 2017 USDA Census of Agriculture, "more than one-quarter of farms have no sales in a typical year, and at least another 30% have positive sales of less than $10,000."

Primary Occupation of Producer: Data on age and primary occupation were obtained from up to four producers per farm. The primary occupation classifications used were:

Farming or ranch work. The operator spent 50% or more of his/her worktime during 2017 at farming or ranching.

Other. The operator spent less than 50% of his/her worktime during 2017 in farming or ranching.

Producer: The term producer designates a person who is involved in making decisions for the farm operation. Decisions may include decisions about such things as planting, harvesting, livestock management, and marketing. The producer may be the owner, a member of the owner's household, a hired manager, a tenant, a renter, or a sharecropper. If a person rents land to others or has land worked on shares by others, he/she is considered the producer only of the land which is retained for his/her own operation. The census collected information on the total number of male producers, the total number of female producers, and demographic information for up to four producers per farm.

Producer Characteristics: Producers (up to four producers per farm) were asked to report primary occupation, sex, age, race, if they were of Hispanic, Latino, or Spanish origin, place of residence, if retired from farming, number of days worked off farm, year in which his/her operation of the farm began, year began operating any farm, if they were a hired manager, if they had military service, and the number of persons living in their households. In addition, the total number of male and female producers was collected from each operation.

Producers of Hispanic, Latino, or Spanish origin: This category is relabeled from 2012.

Producers with Military Service: This category is new for 2017. A producer with military service is a person who currently or previously served on active duty in the United States Armed Forces.

Socially Disadvantaged Farmers and Ranchers: A socially disadvantaged farmer or rancher belongs to a "socially disadvantaged group," which is any group "whose members have been subjected to racial, ethnic, or gender prejudice because of their identity as members of a group without regard to their individual qualities." USDA program agencies differ in their interpretation of this definition, and thus identify somewhat different populations as belonging to socially disadvantaged groups. For example, the USDA's Farm Service Agency defines "socially disadvantaged groups" as including African Americans, Alaska Natives, American Indians, Hispanics (White or otherwise), Asian Americans, and women. The USDA's Natural Resources Conservation Service definition, on the other hand, excludes women from their definition of "socially disadvantaged group." Data presented here follow definitions used by the original data sources.

Race of Producer: Data, except for Hawaii, were collected for American Indian and Alaska Native, Asian, Black or African American, Native Hawaiian and Other Pacific Islander, and White producers. Respondents were asked to mark one or more of the race categories.

Spanish, Hispanic, or Latino Origin: Operators of Spanish, Hispanic, or Latino origin are found in all of the racial groups listed in the census and were tabulated according to the race reported, as well as on tables pertaining only to this group.

Total Income from Farm-Related Sources: This includes gross income from farm-related sources received in 2017 before taxes and expenses from the sales of farm byproducts and other sales and services closely related to the principal functions of the farm business. The data exclude income from employment or business activities, which were separate from the farm business.

Young Producers: A young producer is defined as a producer 35 years of age or younger.

National Overview

General demographic characteristics

Approximately 1.75 million families and individuals and collectively 291,000 partnerships, corporations, and institutional, research, reservation, and other operations own and/or operate approximately 937 million acres of working lands across the United States (USDA 2020). Though land stewards vary in terms of their goals and objectives, shaped by differences in experience level, operation type, and location, in reviewing national data, many patterns emerge. For instance, we find that most landowners are predominantly male (64%; figures 6.2–6.6), non-Hispanic White (approximately 93%), older (80% >45 years), and live on their operations (74%). If we consider operations by acres, 15% of non-Hispanic Whites operate at least 500 acres, and 85% operate at most 499 acres, which are considered small working lands. This smaller working lands pattern is the same across all regions, where at least 83% operate at most 499 acres. Collectively, non-Hispanic Whites operate 93% (847 million acres) of all working lands in the United States, and they will continue to influence the ecosystem benefits derived from their working lands for the near future. Their past, current, and future operational management decisions, including agricultural policy, will continue to influence the prosperity and health of surrounding communities.

Specific demographic and operational characteristics

Many factors continue to inspire novel approaches to land, operation, commodity, and natural resource management and programming, such as age and residence on or off an operation. Aging landowners (>65 years; 34%) control over one-third of all operations and associated lands across the United States. For this reason, we are currently experiencing the largest intergenerational transfer of land-based operations in US history (Bigelow et al. 2016), with implications for future sustainability of operations.

Several demographic characteristics add to the aging landowner challenge: landowner residence, experience, occupation status, and migration patterns away from rural centers. Each is briefly discussed. Landowner *residence* has often been associated with a "tie" to the land or lack thereof. Currently, 26% of landowners live off-farm, also referred to as "absentee landowners." Off-farm residence is often perceived to negatively influence productivity outcomes (Fairchild et al. 2022). This may not be the case in all instances. Both off- and on-farm resident operations contribute to the collective benefits derived from private lands, as numerous land uses yield ecosystem benefits (regardless of residency), and land ownership is an investment, with emotional and financial ties few would willingly depreciate. Residence off-farm has increased over time, and understanding circumstances and motivations behind off-farm residence may help plan for intergenerational land transfers, where those inheriting lands may benefit from off-farm resident landowner technical assistance programs. Agricultural and natural resource professionals should approach landowner support programs by embracing various ownership types and develop targeted stewardship programs to facilitate technical assistance where needed.

Landowner *experience* also may influence intergenerational land transfers. Presently, 75% of landowners have greater than 10 years of farm or ranch operation experience, meaning a quarter of all landowners currently have less than 10 years of experience. It is believed that as aging landowners retire, they will be replaced by a less experienced cohort (Carolan 2018). These cohorts encompass *young* and *new and beginning* landowners, who by default may not have many years of experience in farming. In addition to their limited technical experience, young landowners, by default, lack investment capital to sustain inherited operations. *New and beginning landowners* are typically older, and they may not have the technical experience to operate a farm. They may have gained capital from previous employment or careers to invest in farm operations, but farming requires large capital investments. This group also would benefit from technical assistance programs to begin or maintain farming operations. In terms of intergenerational land transfers, both groups would replace older producers, and they will require resources and training to maintain lands in farming.

Occupation status

This refers to whether farming is a landowner's primary occupation or whether the landowner is employed off-farm as well (farming is not the primary occupation). In comparing landowners nationally, approximately 60% are employed on-farm, meaning farming is their primary occupation or source of income, and 40% report farming is not their primary occupation. As aging landowners are replaced with less experienced landowners, many new landowners are already involved in nonfarming professions. They may elect to remain employed off-farm and balance part-time farming. Other landowners choose to work off-farm to supplement their farming income. There are many reasons that lead to off-farm employment, and understanding these motivations will strengthen technical assistance programming not only for landowners whose primary employment is off-farm but also for young and new and beginning landowner cohorts, who may not want to work solely on-farm but are inheriting land or beginning a farm enterprise. Facilitating technical assistance and resources to these groups as they transition into farming would be beneficial.

The aging landowner challenge is further compounded by *migration* patterns from rural to urban communities, as many potential next-generation landowners have migrated from farm or ranch operations to pursue alternative career opportunities that are more abundant in urban centers. Texas, where 17% of the population lives in rural areas and 83% in urban communities (United States Census 2010), illustrates well the common rural to urban population movement patterns; over the last 50 years, its rural population has migrated to urban communities. These migration patterns and general population trends are not unique to Texas. Consequently, the next generation of working land owners may not necessarily be farm- or ranch-experienced family members, as they have lost both their ties to the land and relevant experience in working the land. Cohorts who inherit lands may not be interested in farming or in returning to rural communities. Selling or renting their inherited operations or subdividing their land influences the nature benefits their operations provide.

Land-use changes influence the ecosystem services derived from rural working lands. Impermeable lands do not yield optimal ecosystem benefits, and impermeable lands do not happen in a vacuum. They are conscious decisions made by landowners and land managers for a variety of reasons (market influences, family or business changes in direction, etc.). Operation lands are divided, sold, or expanded. Buyers and their land-use continuity or changes influence whether land remains permeable or is altered to impermeable uses. This is an area to ponder and explore, for the sake of improving ecosystem service benefits to surrounding communities, as agricultural and natural resource programs prepare for intergenerational land transfers.

Collectively, these estimates illustrate a growing challenge for agricultural and natural resource business segments. Without sufficient strategic and deliberate measures in place to retain operational and land management knowledge and experience, we will be ill prepared to address upcoming intergenerational land transfers. The loss of information, knowledge, and experience captured from years working on the land is a concern for state and federal agencies and landowner advocacy groups who collaboratively work toward facilitating the transfer of information to future land stewards. Looking toward the future, they have developed several "transition" programs and support networks, such as the American Farmland Trust's Farmland for the Next Generation and USDA's Beginning Farmer and Rancher Programs. These collective efforts offer newer landowners access to relevant operation management information beneficial to the ecosystem services influencing surrounding communities.

Associated with information transfer is internet connectivity. Approximately 72% of rural Americans have access to the internet (Pew Research Center 2021). Broadband internet, including satellite, DSL (Digital Subscriber Line), cable, and fiber-optic internet connection types make up 63% of current rural landowner connectivity, and mobile devices comprise another 32% (United States Census Bureau 2010). An increase in broadband adoption has been observed over the last several years and facilitated teleworking as a business model, along with a slow residential shift from urban and suburban communities to rural areas, benefiting the rural community tax base. Although this may stimulate a slight rural land-use change to nonagricultural uses, rural areas benefit from improved connectivity, as they lag behind in reliable internet connectivity when compared to urban communities. Connectivity is important for rural prosperity, quality health care, advanced educational opportunities, and access to increased use of advanced agricultural and natural resource technologies.

What does it all mean?

Demographic trends influence rural areas and agricultural enterprises. Operational decisions, the aging landowner, land-use changes, operation size, and lower income streams, among other factors, amplify the challenges of land stewardship and opportunities for improving ecosystem benefits long term. Figures 6.7 and 6.8 summarize demographic characteristics relevant to program development and the interplay of these variables within communities and operations. Transition programs should consider landowner demographic characteristics in their overall ecosystem service planning and program development, particularly for smaller acreage landowners. Introducing diverse or niche production strategies and sustainable practices should be encouraged on smaller operations to improve profitability for producers, while also assisting the needs of larger, more traditional operations with the goal of creating sustainable operations. This means that land decisions at the operational level are connected to communities near and far and should be considered *one landscape* we should strive to steward.

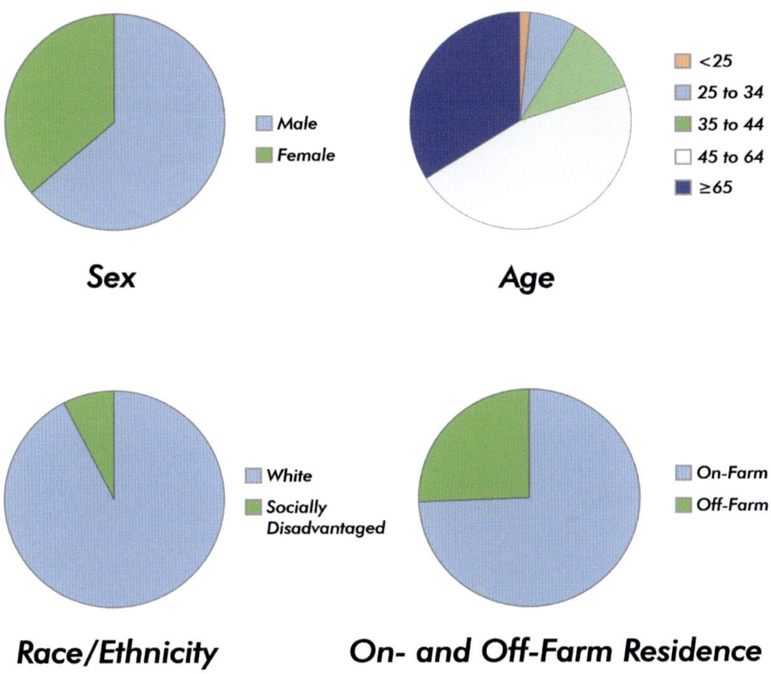

FIGURE 6.2. Sex, age, race/ethnicity, and on- and off-farm characteristics of US landowners, 2017.

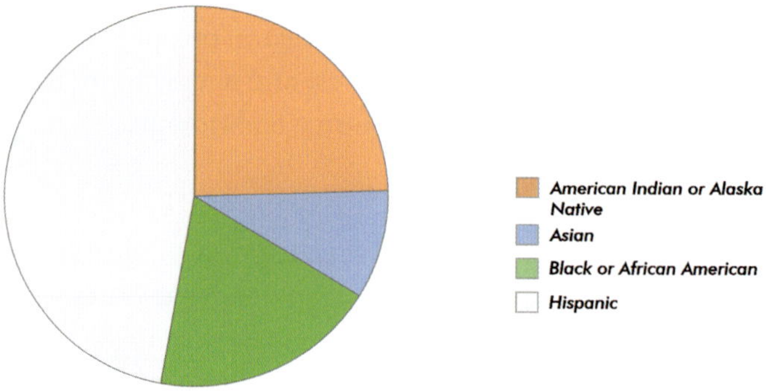

FIGURE 6.3. Socially disadvantaged landowners in the United States, 2017.

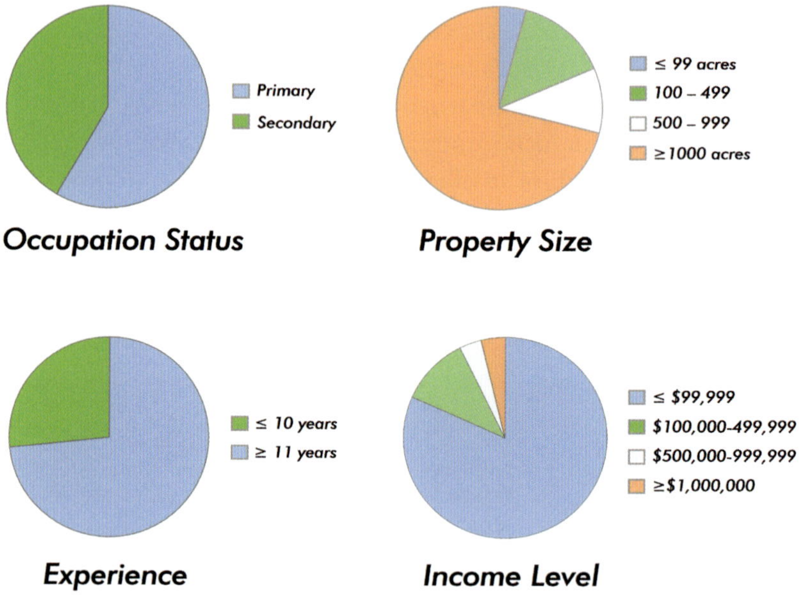

FIGURE 6.4. Occupation status, operations by farm size, experience, and income characteristics of US landowners, 2017.

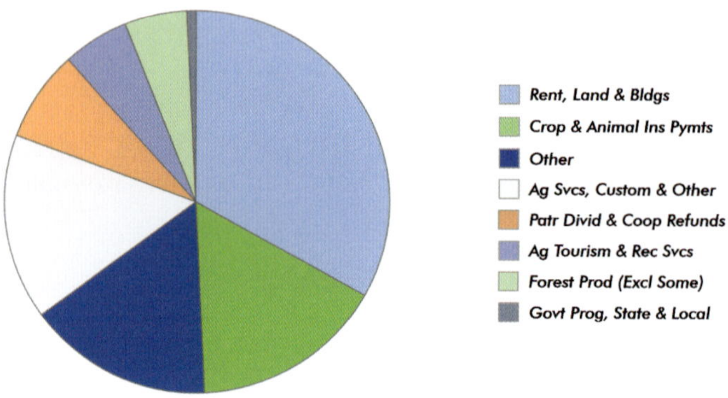

FIGURE 6.5. Farm-related income by receipts in the United States, 2017.

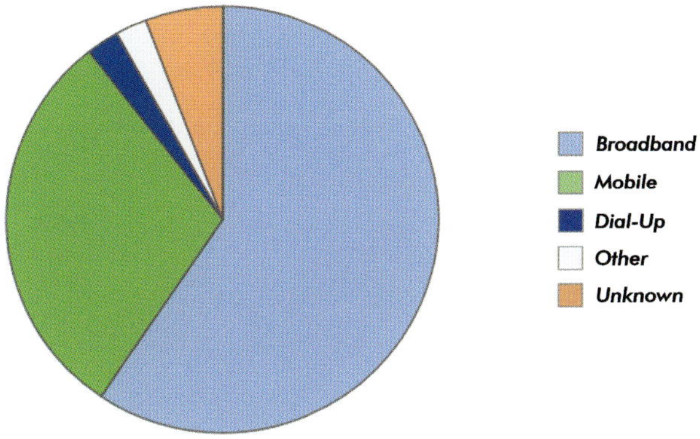

FIGURE 6.6. Landowner internet resources in the United States, 2017.

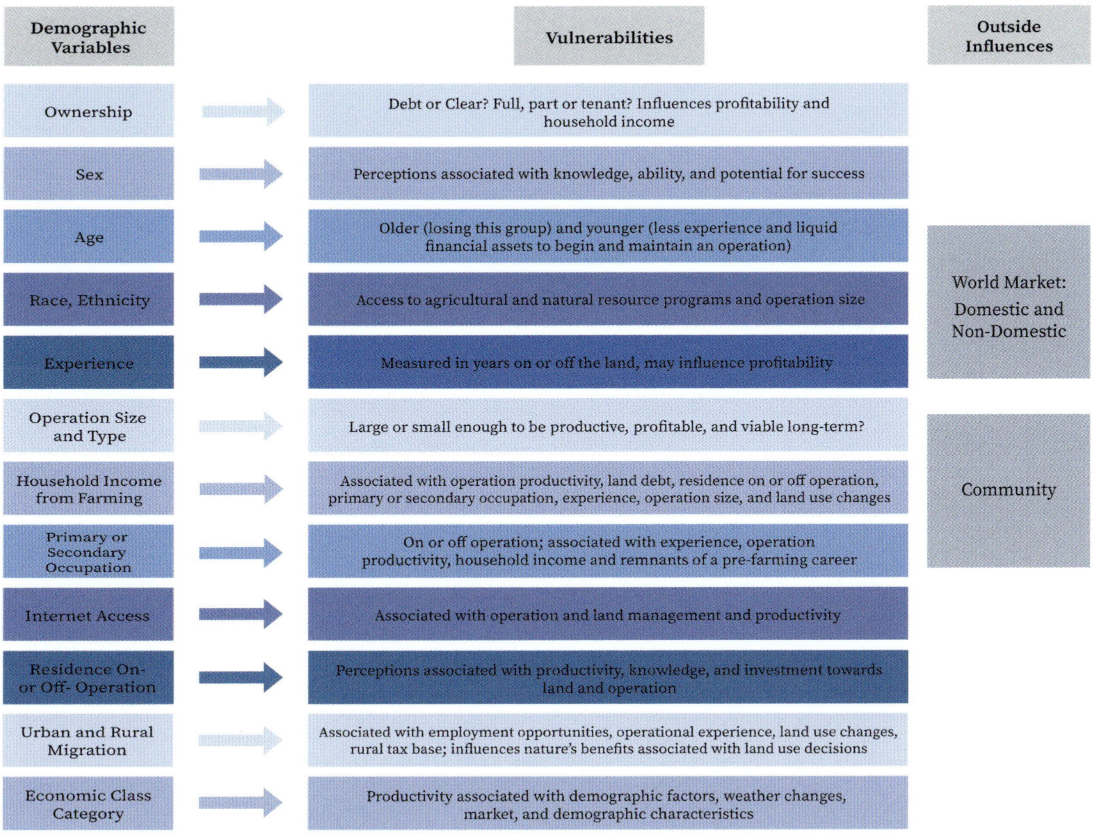

FIGURE 6.7. Landowner demographic characteristics and vulnerabilities to consider in program development, 2023.

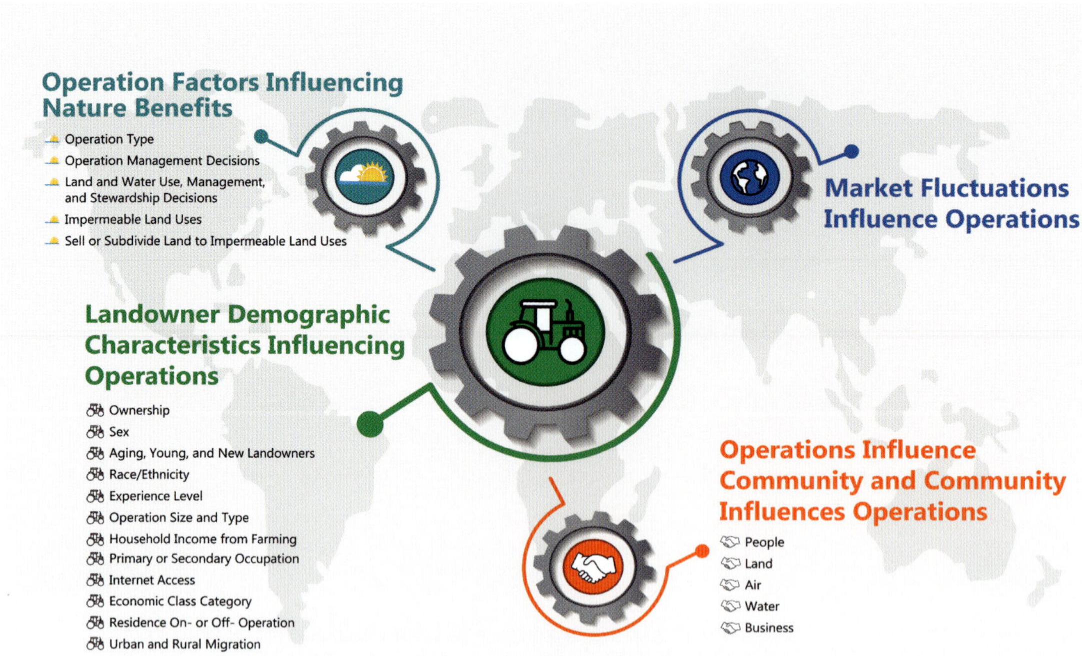

FIGURE 6.8. Interrelationships among demographic characteristics, nature benefits, market fluctuations, and community to consider in program development for landowners in the United States, 2023.

Regional Overview

General demographic characteristics

In addition to our review of landowner characteristics across the United States, we find important regional patterns that may be useful for agricultural and natural resource professionals to understand when creating targeted programs and services. Each demographic variable includes a brief national overview, providing context and relevance, followed by a regional description and comparison. More than one demographic variable is discussed in some sections, as many are interrelated.

Despite their relative uniqueness and independence, operations across the United States are influenced by and susceptible to many factors at the regional level. This section discusses programmatically valuable demographic variables: ownership, internet resources, sex, age, experience, primary or secondary occupation, residence, and military service. Each factor, to a degree, may influence the viability of operations. For this reason, we compared each demographic variable with operation/farm size (minimum-size parcels are generally associated with economic success) and economic class/status to provide information useful in program development. Below is a summary of how information is presented for the regional landowner demographic section:

1. *Specific demographic characteristic:* Ownership, household income, internet resources, sex, age, race and ethnicity, military service, experience, residence, and primary occupation.
2. *Farm/farming label:* For simplicity and ease in presentation, the term *farm* or *farming* is used in the following maps and figures to describe broad agricultural land uses, including ranches, family forests, and wildlife operations.
3. *Map interpretation:* Each variable is described in a *set* containing three maps. *Map 1* includes the variable *by region*. *Map 2* includes the variable *by farm size and region*. *Map 3* includes the variable *by economic status and region*. Below is a description to assist you with interpreting the maps.
   a. Each map contains *pie charts* within each region. A pie chart *color legend* describes the categories for each variable. For example, the variable *sex* contains two categories, *male* and *female*, so the pie chart legend contains two colors. The pie chart for the variable *internet resources* contains five categories based on internet type: *broadband, mobile, dial-up, other,* and *unknown*. So pie charts for internet resources contain five colors, one for each internet type.
   b. Each pie chart is divided into *quarters*. Each quarter represents a *farm size* category or an *economic status* category. Farm size categories include ≤99 acres, 100–499 acres, 500–999 acres, and ≥1,000 acres. Economic class or status categories include ≤\$99,999, \$100,000–\$499,999, \$500,000–\$999,999, and ≥\$1,000,000.

Ownership

Land ownership in the United States is both associated with and influenced by several factors, including legacy, the passing of one's landholdings to children or family, also known as inheritance. The USDA describes ownership with the term *tenure*, which describes an operation's ownership as full ownership, partial ownership (partial ownership of operations), or tenancy (renting land from others to produce) of operations (figure 6.9). When comparing ownership types across the United States, partial ownership is greatest in the Northern Plains and Midwest regions.

The full ownership of land-based operations varies among all landowners by race and ethnicity (Asian American, 78%; American Indian or Indigenous, 43%; African American or Black, 66%; Hispanic, 78%; non-Hispanic White, 70%) and by sex (females, 75% of operations and 35% of all female acreage; males, 68% of operations and 34% of all male acreage). Most fully owned operations consist of 10 to 179 acres (females, 59%; males, 56%) and over 180 acres (females, 20%; males, 32%).

Ownership greater than 500 acres also varies among all landowners (Asian American, 5%; American Indian or Indigenous, 17%; African American or Black, 4%; Hispanic, 10%; non-Hispanic White, 15%). Non-Hispanic Whites operate most acreage in the United States (90%; Asian American, 0.25%; American Indian or Indigenous, 6%; African American or Black, 0.44%; Hispanic, 3%). Regardless of race and ethnic characteristics, most US landowners manage smaller operations (<500 acres); conversely, fewer landowners manage most US acreage via larger operations. As we review many factors associated with land ownership, it is relevant to consider race and ethnicity, among other factors, for developing programs that ensure the legacy and heritage of all landowning families, especially those most vulnerable to losing their lands. Ensuring the American spirit of land ownership across cultures and generations provides continuity of benefits derived from the land that are beneficial to communities nationwide.

What does this mean?

Land ownership is discussed throughout this chapter as a reference for managing lands and determining appropriate management and conservation strategy programming. What may be meaningful for larger landholdings may not be relevant for smaller landholdings. The same is true for ownership type (full ownership, partial ownership, and tenancy). Income derived from various types of ownerships also should be considered when developing targeted program strategies. Consider the economic benefits derived from each ownership type via the cost associated with using the land for production: Tenants pay for their land use and subtract land-use fees from their profit; partial owners earn partial incomes from the full land operation; full owners without a lien earn 100% of production income.

Also associated with ownership is an operation's economic class category. Economic class considers the sum of market value of agricultural products sold and federal farm program payments. Both farm size and economic class are good descriptors of operations from a program development

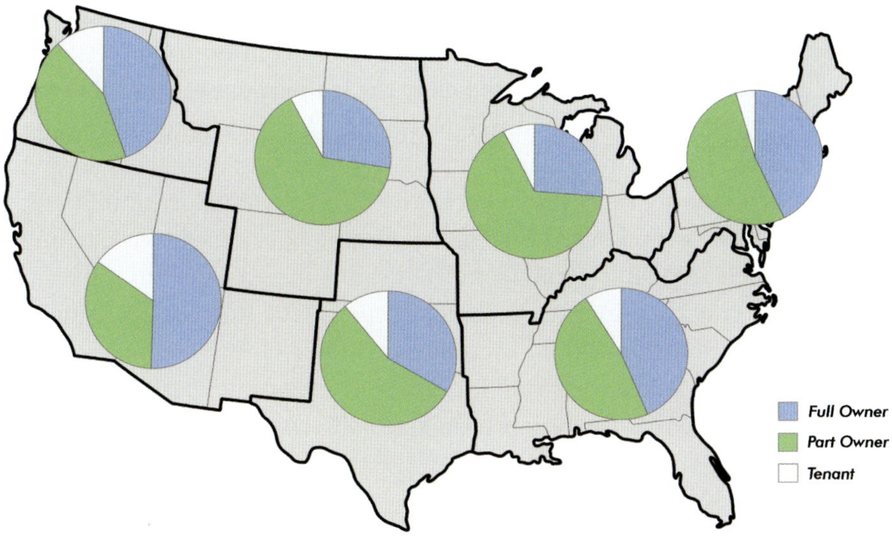

FIGURE 6.9. Ownership characteristics of US landowners, 2017.

perspective; therefore, each variable (age, income, internet service, etc.) will be associated with both farm size and economic class. With some exceptions, the general rule of thumb is that the larger the operation, the larger the economic class category (i.e., the more income the landowners earn from their land-based operations). The strategies landowners employ in managing their lands and agricultural commodities are associated with their income.

Household income from farming

This is a measure of earnings a given household derives annually from their land. Three quarters of all US farming households derive wages from off-farm employment, so general household income by itself may be deceiving; most landowners earn <$100,000 annually from their operations regardless of race or ethnicity. Operation size is often associated with household income from farming, and most farmlands comprise 499 acres or less (<50 acres, 25% of all farming households; 50–499 acres, 50% of all farming households). At a national level, 70% of all farming households earn less than a quarter of their annual income from their lands, while approximately 13% derive at least three quarters of their income from their lands. This makes sense because most farming households generally consist of smaller parcels and off-farm employment is common. It is assumed that as economic class categories increase, so would household income from farming, but the relationship among the two factors may not be direct; otherwise, off-farm employment would not be as common.

At a regional level, the northern half of the United States (Pacific Northwest, Northern Plains, Midwest, and Northeast) derives a larger percentage of household income from farms consisting of at least 500 acres. Larger income brackets are more prevalent in the Midwest and Northern Plains; however, across the United States, the number of households earning 100% of their annual income from their lands increases as their operation's economic class increases. The opposite is true for those earning at most 25% of their household income from their lands. The number of these households progressively increases as economic class categories decrease. Comparing operation farm size and economic class across the United States, we find smaller size classes (99 acres or less) consist mostly of smaller economic classes.

Approximately half of the Southwest's largest farms (1,000 acres or more) consist of operations within the $99,999 or less economic class category. However, in the Midwest, farms of 500–999 acres have proportionally more operations within the $100,000 or more economic class categories. As farm size increases to 1,000 acres or more, this region also has the largest proportion of operations within the $500,000 or more economic categories, across all US regions. Larger size classes appear associated with larger economic classes, except for niche agricultural products (figures 6.10–6.13).

What does this mean?

Household income from farming may be directly associated with off-farm employment. If we follow the logic that economic class increases as operation size increases, then operations with off-farm and on-farm resident landowners would appear equally profitable, particularly off-farm resident operations, which are common among larger operation size classes, where it would appear more difficult to manage operations well (see the *Residence* discussion for more information). Data in this chapter do not necessarily support the assumption that operations with off-farm resident landowners are less profitable or less well managed. More research is needed in this area to better inform outreach programming. Data do support the assertion that household income from off-farm employment is assisting farming households by supplementing their incomes. Opportunities may exist to incorporate incentives or programs that may boost farming household income derived from lands. These may include conservation-based incentives, among others. From a landscape perspective, all lands influence nearby communities, and incentivizing community-beneficial stewardship practices may improve farming household income and the natural benefits derived from lands.

Internet resources

Internet resources are necessary for the success of most business enterprises. From receiving and communicating ideas to downloading resources and data for direct application on land-based operations, internet access improves businesses and assists communities. Internet infrastructure is divided

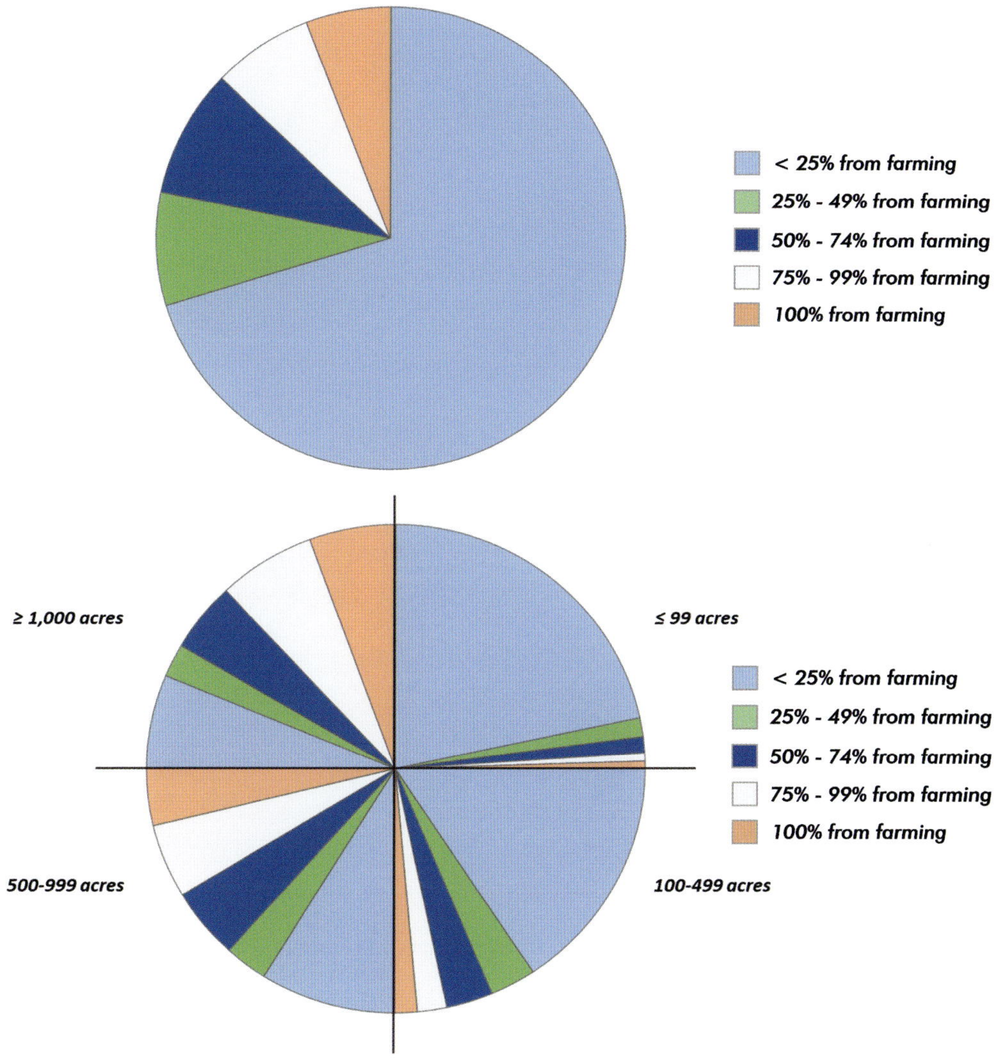

FIGURE 6.10. Percentage of household income from farming (overview) and by farm size in the United States, 2012.

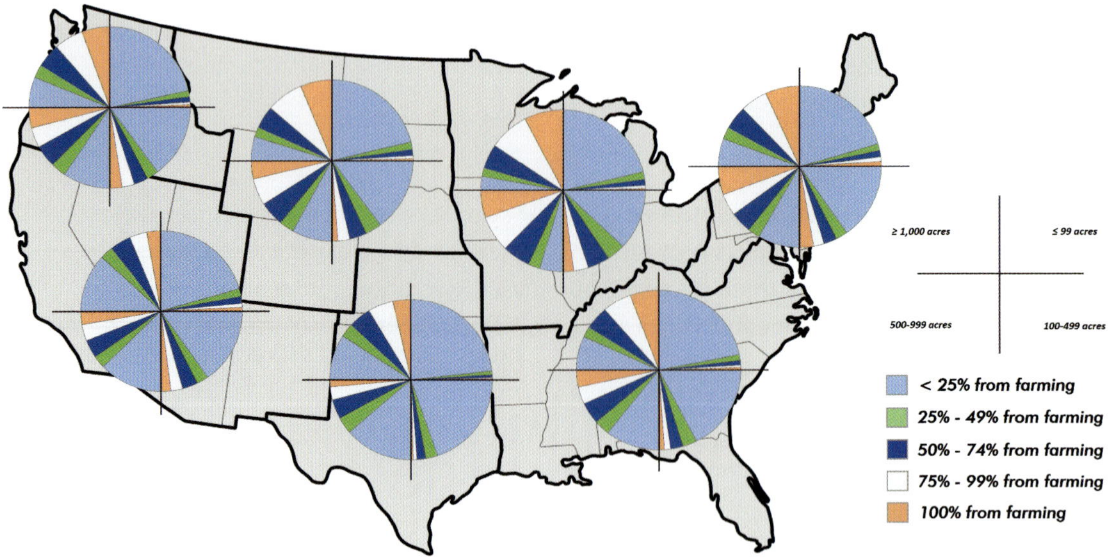

FIGURE 6.11. Percentage of household income from farming and farm size for landowners in the United States, 2012.

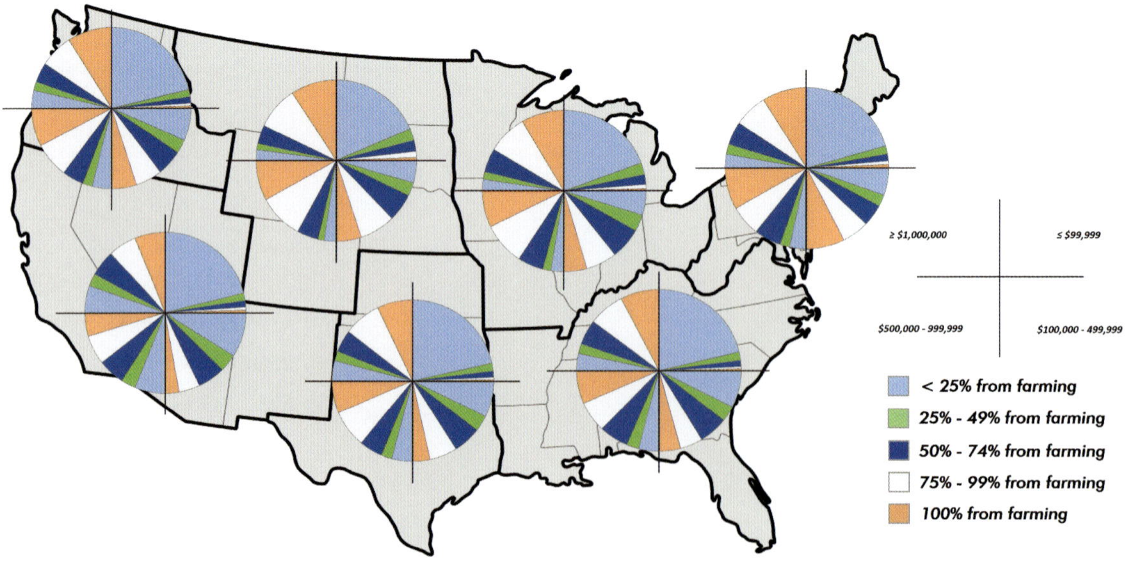

FIGURE 6.12. Percentage of household income from farming and economic status for landowners in the United States, 2012.

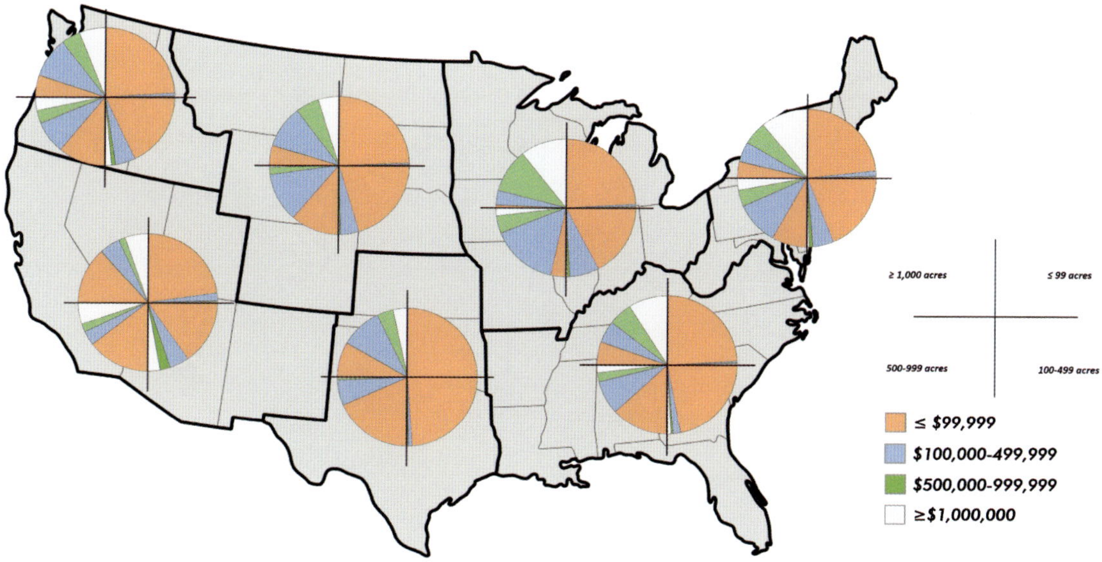

FIGURE 6.13. Farm size and economic class by region, 2017.

into several types, with broadband, mobile, and dial-up as the most common among US landowners. There is a push by the US government to improve internet access across rural regions, and a focus on download/upload speed is consistent with the improvements. In all seven regions, there are some areas where the internet type is unknown (5%–7%) and where dial-up internet is still used (2%–3%; figures 6.14–6.16). Collectively, these encompass 7%–10% of all operations per region. There is no significant difference regionally when comparing internet type with economic class and farm size categories. This is not surprising as internet availability in communities is not influenced by landowners. The near uniformity of internet types across the United States indicates similar internet availability and perhaps an upgrade from older technologies (dial-up) to more modern internet access, such as broadband and mobile phones. Although internet availability and type appear uniform across the United States, the data resolution offers a high-level overview of demographic trends. Differences in internet access, such as download speed and time, may exist within regions, resulting in differences in accessibility of resources.

Areas for improvement include determining internet type in regions where internet type or availability is "unknown" to ensure uniform access. This would also aid in the assessment of how the internet is used within farming households to then determine the feasibility and adoptability of

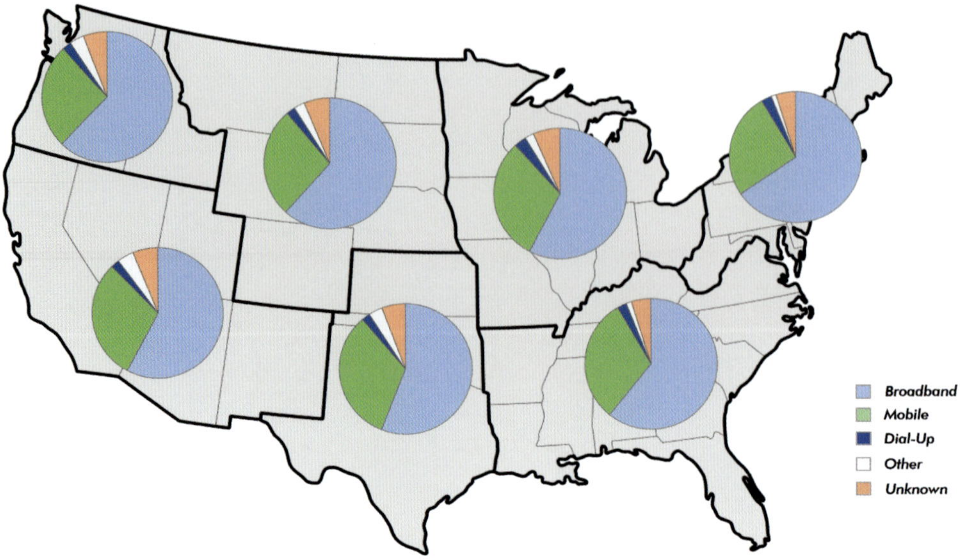

FIGURE 6.14. Landowner internet resources in the United States, 2017.

FIGURE 6.15. Landowner internet resources and farm size by region, 2017.

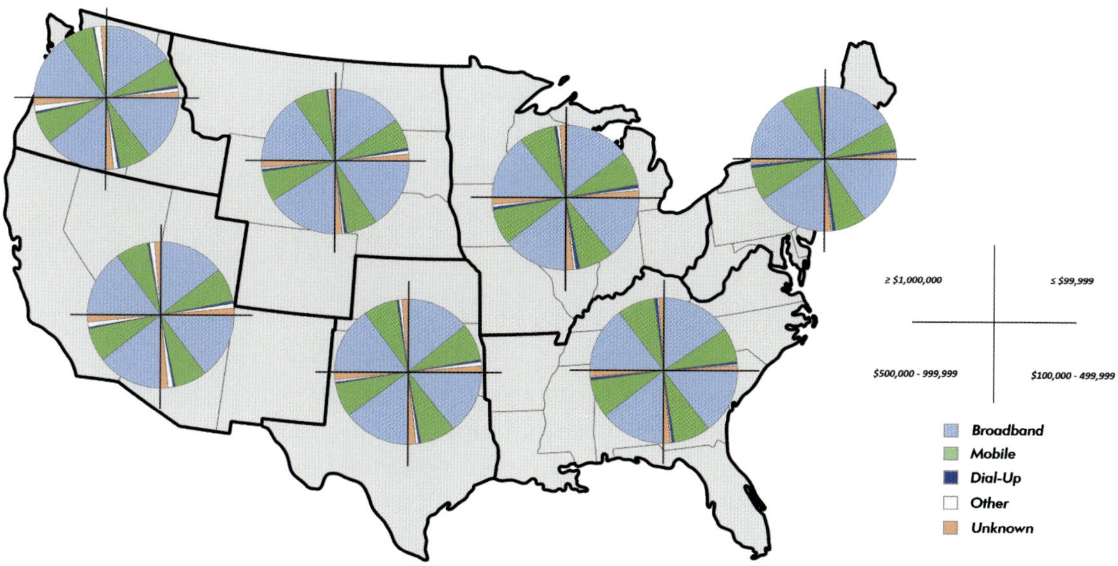

FIGURE 6.16. Landowner internet resources and operation economic class by region, 2017.

precision agricultural technologies, indicate the best method for communicating agricultural and natural resource–related information, and provide data on how often informational resources are accessed to better manage operations and land resources. Assessing communication strategies is important; however, a focus on reaching landowners and fostering meaningful connections, determining their practical needs, and finding ways to help them balance off-farm employment with on-farm demands would be more helpful than academic exercises on communication.

What does this mean?

Information and communication are keys to effectively manage lands and operations, and meeting the internet needs of landowners is vital to landowner success. Education and outreach providers may create products such as videos on farming or natural resource management practices that are beneficial, but landowners without proper internet access or knowledge will not be able to utilize those resources. It would benefit all parties to emphasize internet access, reliability, and use by farming households and to establish communication strategies to better understand landowner information needs and preferences that may be appropriately met by informational products via internet services.

Sex

Approximately 64% of private landowners in the United States are male and 36% are female, with no regional variation. Throughout the United States, the percentage of female landowners decreases as farm size and economic class categories increase. The only exception is the Southeast, where the proportion of female landowners remains relatively equal as economic class increases (figures 6.17–6.19).

What does this mean?

Given that female landowners are more likely to operate smaller operations, which are associated with smaller economic class categories, they also may be deriving less income from those operations or seeking more off-farm employment to supplement their household income than their male counterparts. Programs that target small-acreage landowners should be cognizant of this factor and adapt programming to consider the needs of female landowners or create female-centered programming. Targeted, incentive-driven programming focused on landowners of smaller farm size categories or female landowners may be beneficial to help them supplement their household income from farming.

Age

The average age of all landowners in the United States is approximately 57 years. Most landowners are 45 years or older (80%). Few landowners are 44 to 35 years (12%), 34 to 25 years (7%), and younger than 25 years (2%). Regional differences were noted in operation size and economic class categories across all age groups. While the youngest three age groups are less prevalent across all farm size and economic class categories, landowners younger than 25 years were significantly less prevalent in all operation size and economic class categories, making up 1% or less in each, except in the Northeast, where they made up between 2% and 3% (figures 6.20–6.22). Although landowners younger than 25 years may not commonly have access to capital for investing in land-based operations, they are a crucial group for replacing older landowning generations. See Bigelow et al. (2016) for ownership, tenure, and transfer trends.

Despite an increase in accessibility to capital for landowners between the ages of 25 and 44 years, few are selecting farming as a career. Farming requires large investments, and it is understandable that younger age groups may have difficulty starting agricultural enterprises. Intergenerational land transfers and land programs might be an option for those seeking a farming profession. Renting land is also common among younger age

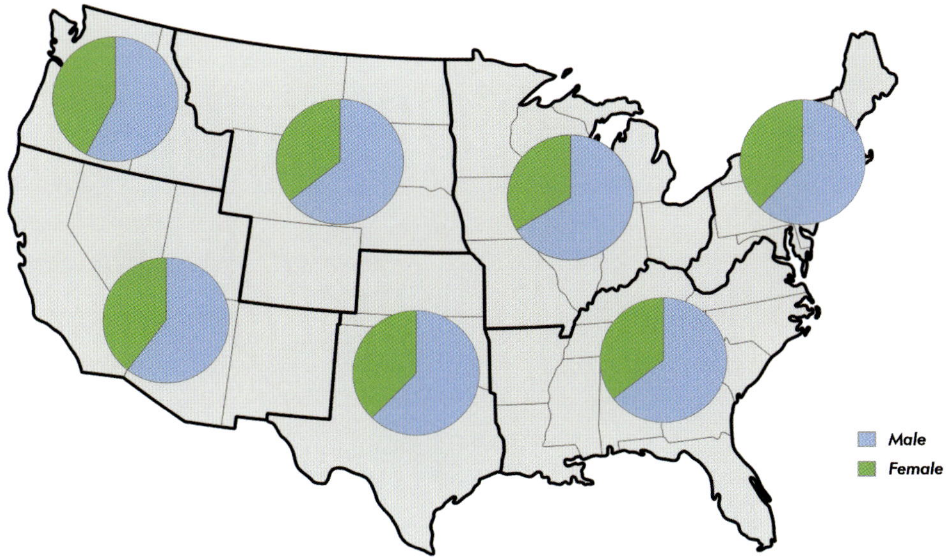

FIGURE 6.17. Landowners by sex and region, 2017.

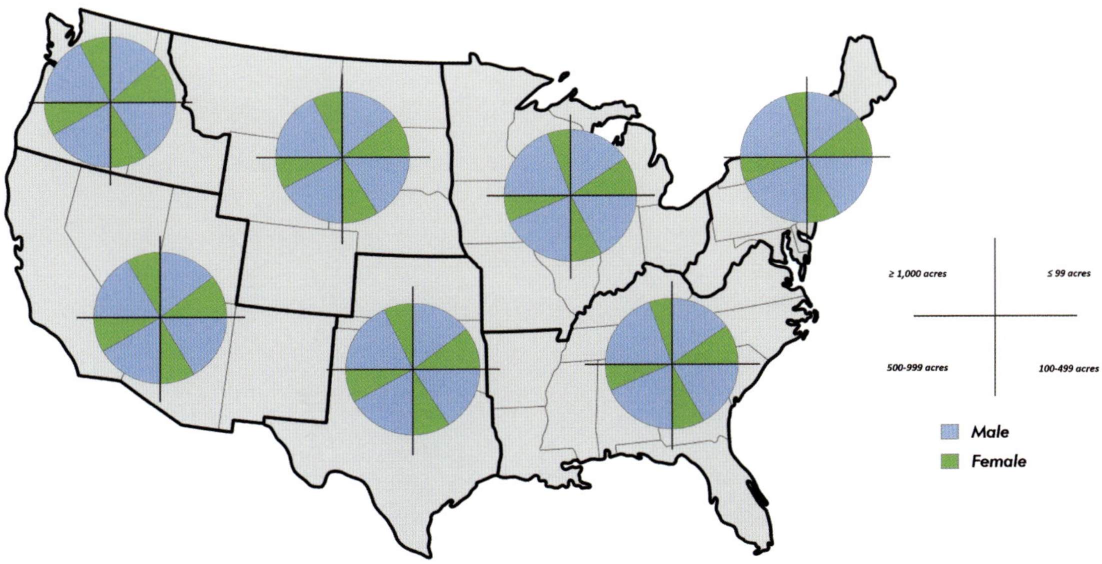

FIGURE 6.18. Landowners by sex, operation size, and region, 2017.

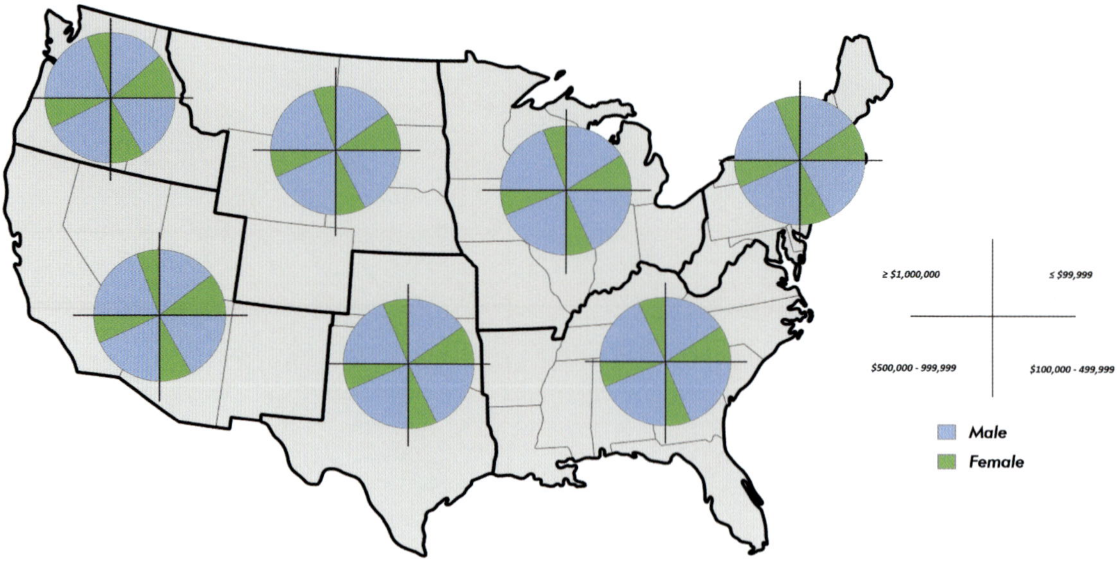

FIGURE 6.19. Landowners by sex, economic class, and region, 2017.

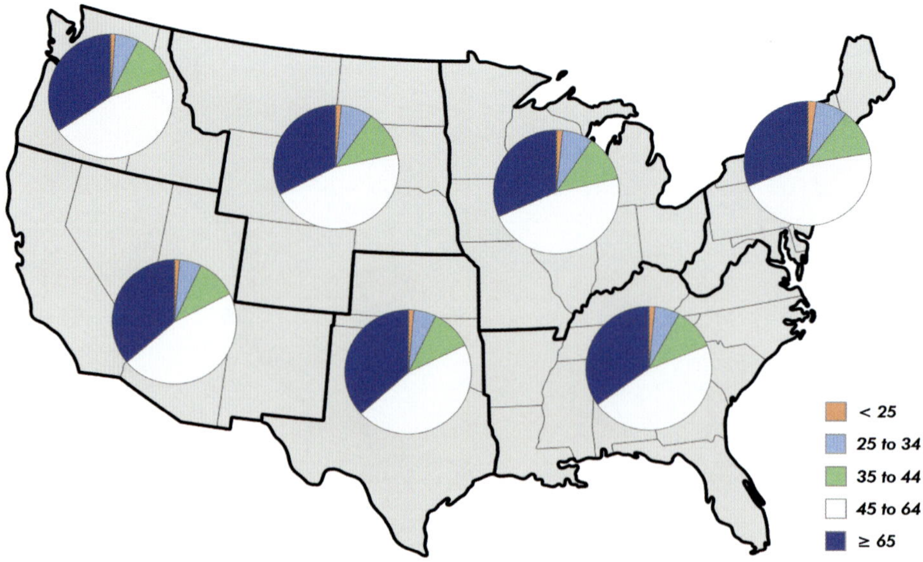

FIGURE 6.20. Landowners by age (in years) and region, 2017.

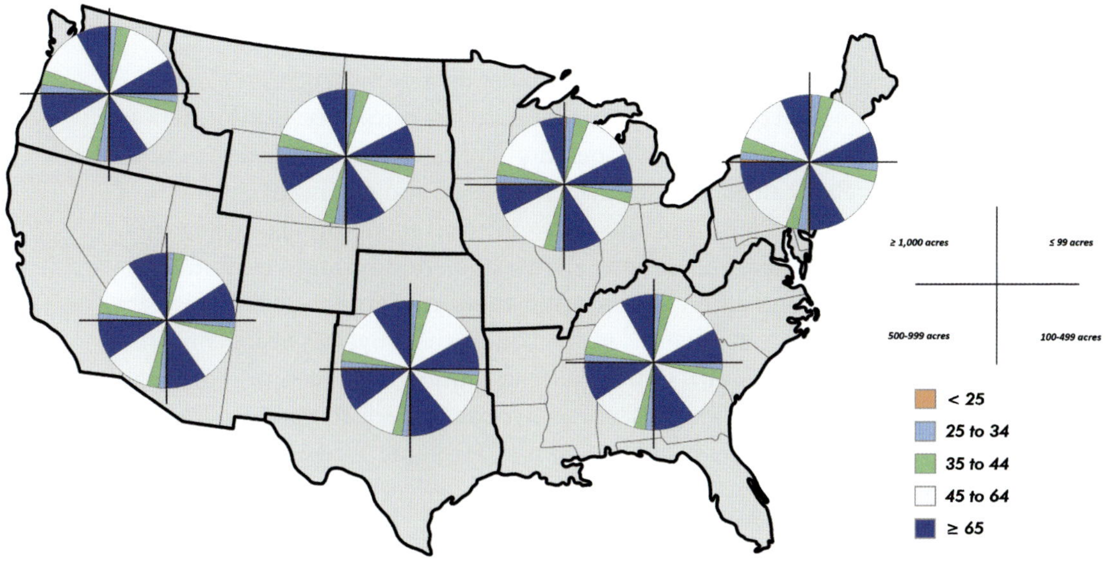

FIGURE 6.21. Landowners by age (in years), operation size, and region, 2017.

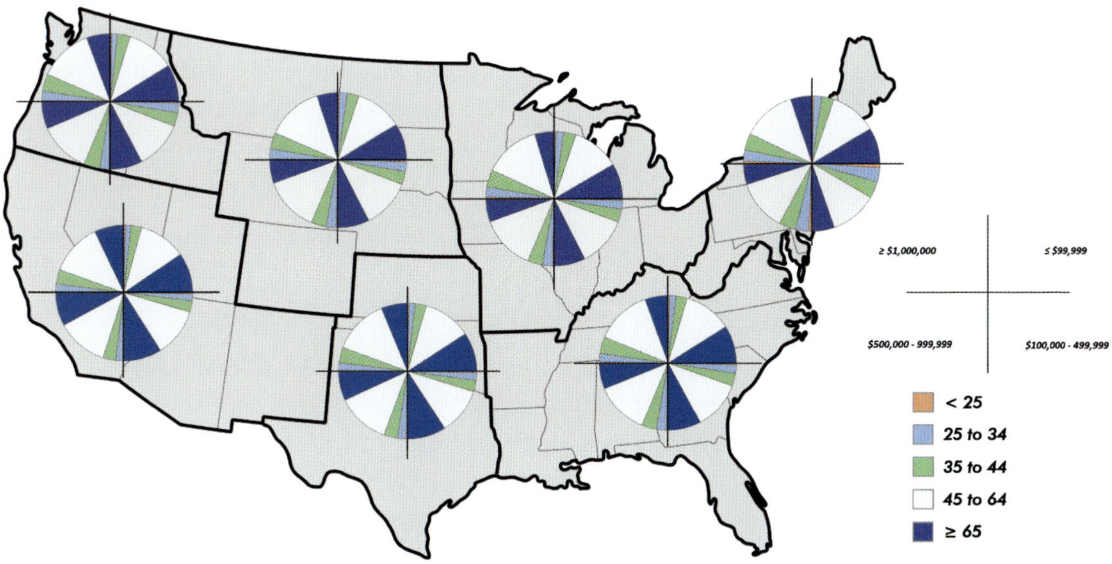

FIGURE 6.22. Landowners by age (in years), economic class, and region, 2017.

groups, as it helps them gain experience in farming and access to capital for purchasing land.

Generally, landowners 45 years and older operate most acreage across the United States and the majority of economic class categories: Landowners 45 to 64 years own 42%–58% of all operation size categories and 44%–54% of all economic class categories; landowners 65 years and older own 23%–42% of all operation size categories and 19%–39% of all economic class categories.

When we compare age among all race and ethnic groups, we find 46%–57% of all race and ethnic group landowners are between 45 and 64 years old, and 34%–46% are over 65 years. Landowners over 45 years comprise approximately 86% of Asian American landowners, 88% of American Indian landowners, 93% of Black or African American landowners, 86% of Hispanic landowners, and 89% of non-Hispanic White landowners.

What does this mean?

Approximately 80% of US landowners are 45 years of age or older, and of these, 46% are between 45 and 64 years old, and 34% are 65 years of age or older. These landowners are nearing or have surpassed retirement age, suggesting approximately one-third of US agricultural operations are potentially in an intergenerational land ownership transition or will be soon.

The growth and success of the youngest age groups will shape the future of US agriculture as the country modernizes agricultural practices and prepares for a second, larger intergenerational land transfer of landowners who are currently between the ages of 45 and 64 (46% of all US landowners).

Race and ethnicity

The term *socially disadvantaged* is used by USDA to collectively describe landowners of Hispanic, Black or African American, Asian American, Indigenous or Native American, and Hawaiian or Pacific Islander descent. For purposes of this chapter, Hawaiian or Pacific Islander and Alaskan communities were not considered in the analysis, which focused on the 48 contiguous states. Historical collective decisions are likely associated with the percentage of socially disadvantaged landowners, who are represented in all operation size and economic class categories in varying percentages.

Across the United States, most landowners are non-Hispanic White (93%; average age, 57.5 years; operate <500 acres, 95%). In the Midwest, Northeast, Northern Plains, and Pacific Northwest, they comprise 98%, 97%, 96%, and 94% of landowners, respectively (figures 6.23–6.25). The

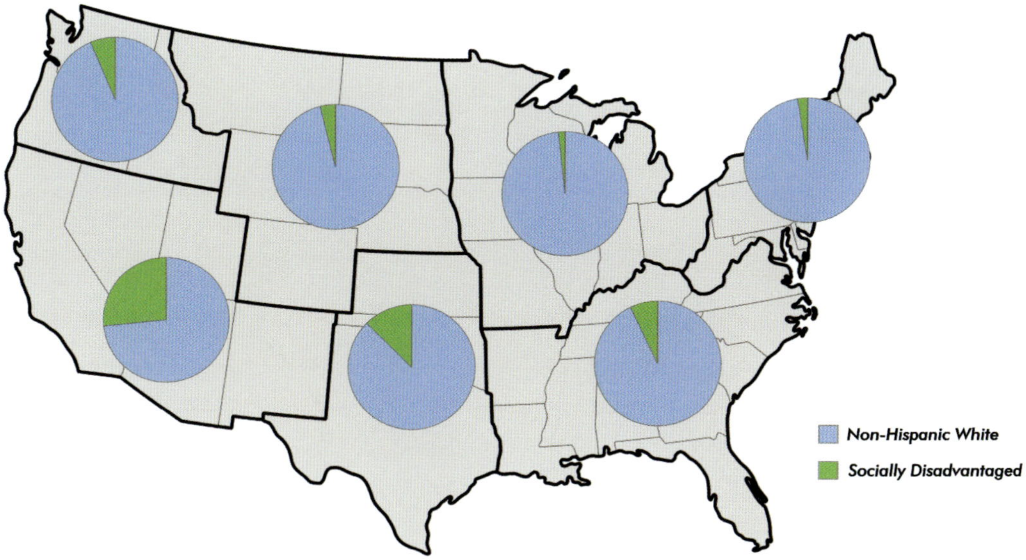

FIGURE 6.23. Landowners by race/ethnicity (non-Hispanic White, Socially Disadvantaged) and region, 2017.

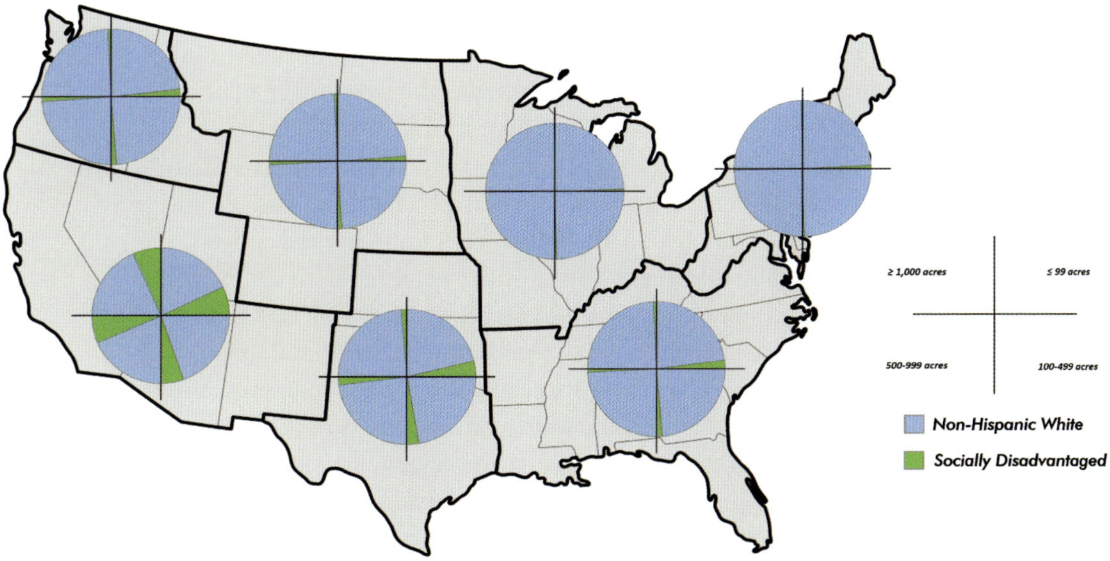

FIGURE 6.24. Landowners by race/ethnicity (Non-Hispanic White, Socially Disadvantaged), operation size, and region, 2017.

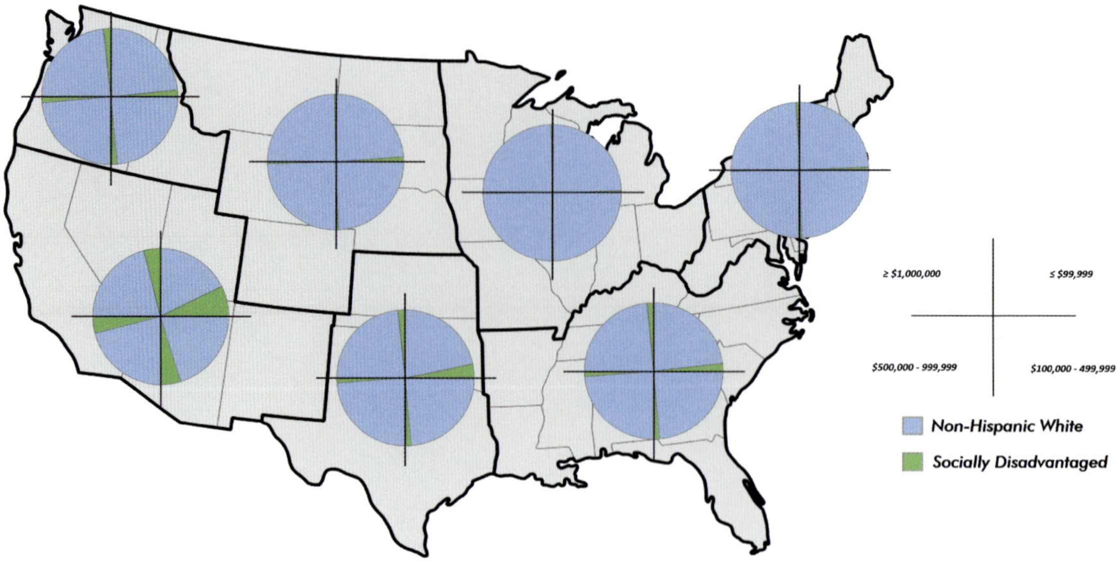

FIGURE 6.25. Landowners by race/ethnicity (Non-Hispanic White, Socially Disadvantaged), economic class, and region, 2017.

percentage of socially disadvantaged landowners increases slightly in the lower half of the United States (Southwest, 27%; Southern Plains, 13%; Southeast, 7%). Except for the Southwest, the percentage of socially disadvantaged landowners decreases as operation size and economic class categories increase (1% to 10% change); the percentage of non-Hispanic White landowners increases as operation size and economic class categories increase (1% to 13% change). In the Southwest, with few exceptions, the percentage of socially disadvantaged groups remains relatively the same across all operation size and economic class categories (operation size: 99 acres or less, 28%; 100–499 acres, 21%; 500–999 acres, 26%; and 1,000 acres or more, 27%; economic class categories: $99,999 or less, 29%; $100,000 to $499,999, 19%; $500,000 to $999,999, 16%; and $1,000,000 or more, 16%).

Few socially disadvantaged groups own and operate farms in the United States (American Indian or Indigenous landowners: comprise 2% of all landowners, average age, 57 years, operate <500 acres, 83%; Asian American: comprise 1% of all landowners, average age, 56 years, operate <500 acres, 95%; African American or Black: comprise 1% of all landowners, average age, 61 years, operate <500 acres, 96%; Hispanic: comprise 3% of all landowners, average age, 55 years, operate <500 acres, 91%).

Understanding socially disadvantaged landowner needs and helping them succeed in their land-based enterprises are beneficial to all, as their success improves the heritage and legacy of farming in an era when maintaining land in farms becomes increasingly challenging. Developing targeted programs that focus on the operational success of socially disadvantaged landowners and making precision agriculture technologies and meaningful technical and incentive programs accessible are both crucial to the longevity of their farming practices, legacy, and heritage. These efforts also will increase the natural benefits derived from the land for nearby communities, such as clean air and water, among many others. Every *acre* and every *landowner* matters.

What does this mean?

Although socially disadvantaged landowners are generally represented across operation size and economic class categories, non-Hispanic Whites represent most US landowners; their land-based operations and stewardship practices have and will continue to influence most land and water resources across the country, to include the ecosystem services influencing communities. Despite their small numbers, socially disadvantaged landowners also provide critical nature benefits to communities, and they too would benefit from the vast suite of local, regional, and national programming on novel and updated approaches to agricultural practices, operation management, and land and water resource stewardship. Additionally, because every acre matters, access to targeted agricultural and natural resource programming, such as commodity-based technical assistance programs and outreach, also would be helpful to the operational success, heritage, and legacy of their farming practices and lands. Sample programs that provide technical assistance to landowner communities and foster collaborative networks include USDA's Socially Disadvantaged, Beginning, Limited Resource, and Female Farmers and Ranchers program, Thomas Mitchell's heirs' property efforts, and Kimberly Ratcliff's 100 Ranchers program. Collaborative network development programs foment strong rapport, support, and community landowner learning networks, where idea, information, and experience exchanges are beneficial.

Military service

Approximately 11% of landowners have served in the military (female, 36%; male, 64%), which is slightly higher than the national average (7%), and military service is represented across all operation size and economic class categories. In comparing military service across race and ethnic landowner groups, all have served in the military to varying degrees (Hispanic and non-Hispanic Whites, 11% each; Indigenous or Native Americans, 10%; Asian Americans, 6%), with African American landowners representing the group with the most military service, almost double at 20%. Slightly over half of landowners with military service engaged in farming as a primary occupation (52%), and this varied slightly by race and ethnicity (Asian American, 50%; American Indian or Indigenous, 49%; Black or African American, 44%; Hispanic, 40%; and non-Hispanic White, 41%). When military service and years of experience were compared, 80% of landowners who served had greater than 11 years of farming experience (American Indian or Indigenous, 74%; non-Hispanic White, 73%; Black or African American, 72%; Hispanic, 64%; Asian American, 61%). In contrast, new and beginning landowners with military service are few (7%), and those who are younger than 35 years are fewer still (2%). Most landowners who serve or have served in the US military are 65 years and older (70%). Across US regions, service was near uniform, and the percentage of landowners with military service decreased as operation size and economic class categories increased (figures 6.26–6.28).

What does this mean?

Military service is much appreciated across the United States. According to our findings, military service is primarily associated with older landowners, and very few younger landowners and new and beginning landowners have served in the military. In appreciation for military service, the federal government has established incentives for service members and veterans, yet we find that, uniformly, military service decreased slightly as operation size and economic class categories increased. Landowner military service participation, in general, can be associated with historical events, yet land is a finite resource. Despite historical accessibility to resources, former military personnel do not own larger operations (>500 acres). This depicts either the challenges of owning and managing larger operations, the inaccessibility of resources (financial or informational), or the balancing of off-farm employment. When developing programs for military veterans, access to financial and informational resources should be a consideration. Younger landowner generations do not appear to serve in the military at the same level as their older counterparts; neither do new and beginning landowners, and this also might be considered when developing outreach and engagement strategies.

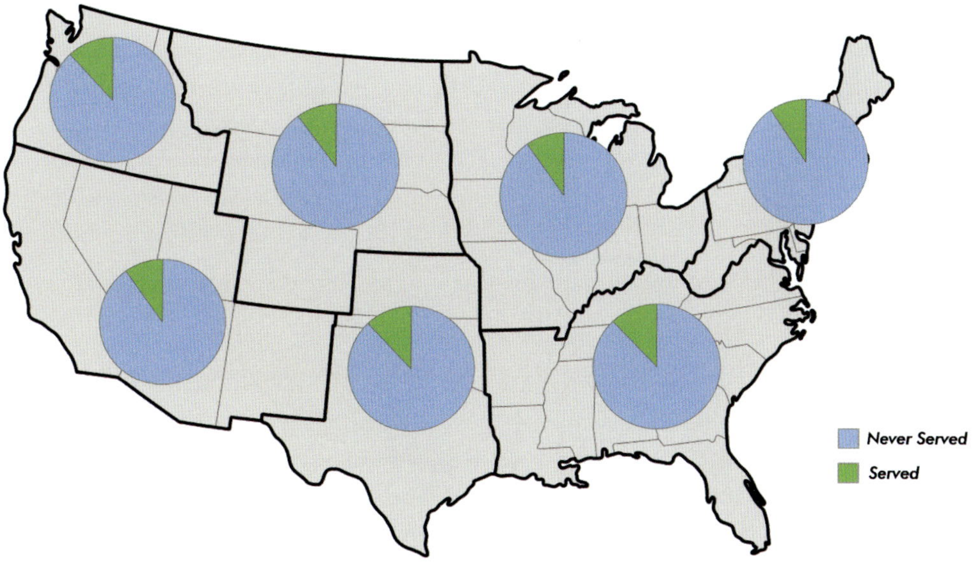

FIGURE 6.26. Landowners who served in the military, by region, 2017.

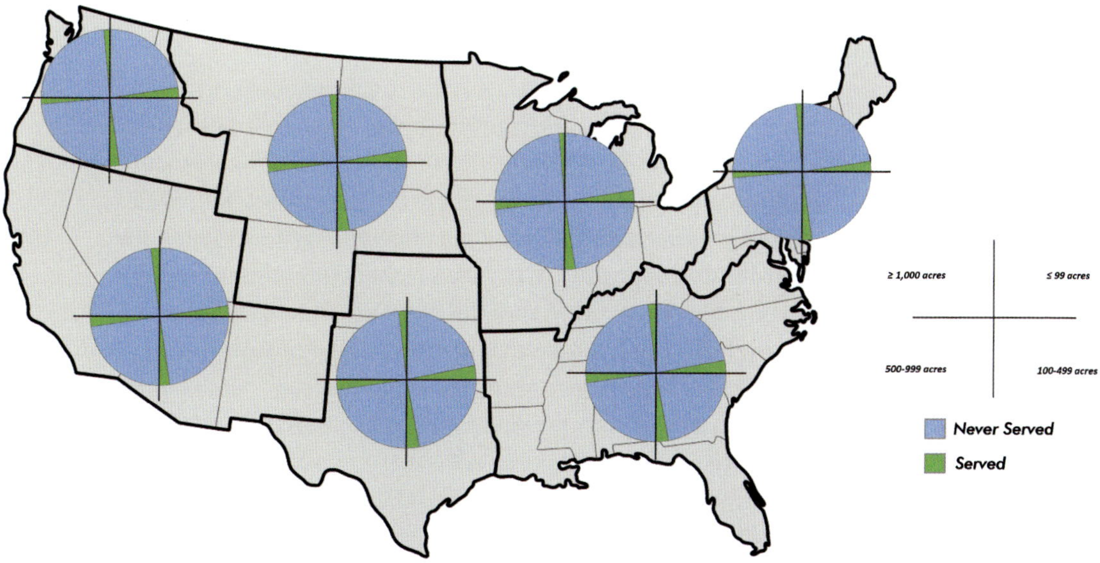

FIGURE 6.27. Landowners who served in the military, operation size, and region, 2017.

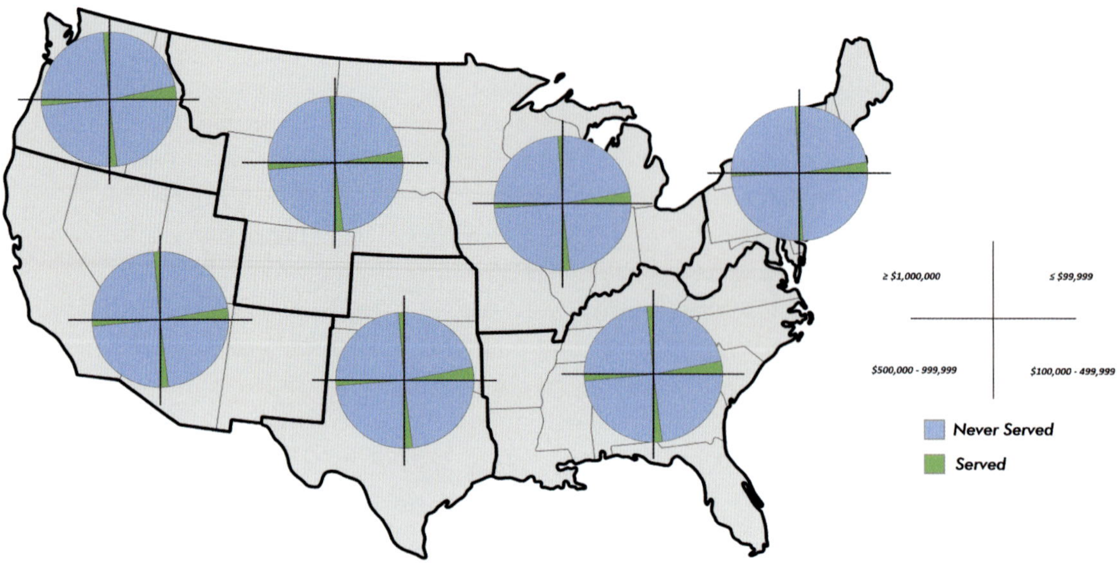

FIGURE 6.28. Landowners who served in the military, economic class, and region, 2017.

Experience

One measure of landowner experience level is based on the number of years worked on any operation. Two significant landowner groups fall within this category: *new and beginning landowners* and *younger landowners*. Both groups are different, very small in number, and each is necessary for replacing aging and retiring, experienced landowners (see the *Age* discussion for more information about this group). This poses a challenge associated with intergenerational land and operation transfer. For this reason and because specific federal programs are in place to assist these two groups, they are discussed in depth here. From a programmatic perspective, establishing strong recruitment and retention programs for these two groups makes sense for the longevity of farming operations.

Landowners with less than 10 years of experience on any operation are considered *new and beginning landowners* (referred to here as *beginners*), and they make up 33% of all landowners with an average age of 46 years (overall) and off-farm residence of 33%. *Beginner* landowner operations are primarily owned by males (59%), and most comprise less than 500 acres (90%). Only 28% of beginner landowners participate in farming as their primary occupation, and they fully own (39%), partially own (47%), or are tenants (14%) of their operations. Beginner landowners represent between 33% and 50% of each landowner group: Asian Americans make up 1% of all beginners, and beginners comprise 51% of all Asian American landowners. Black or African Americans make up 1% of all beginners, and beginners comprise 34% of all Black or African American landowners. American Indians or Indigenous make up 2% of all beginners, and beginners comprise 33% of all American Indian or Indigenous landowners. Hispanics make up 4% of all beginners, and beginners comprise 45% of all Hispanic landowners. Non-Hispanic Whites make up 92% of all beginners, and beginners comprise 33% of all Non-Hispanic White landowners.

Younger landowners—those who are 35 years and younger—comprise 12% of all landowners, with 33% residing off-farm. By default, most younger landowners have less than 10 years of experience operating their land-based operations (80%). Approximately 64% of all young landowners are male, and most (83%) manage less than 500 acres. Only 35% of all younger landowners participate in farming as their primary occupation, and they may fully own (54%), partially own (30%), or are tenants (16%) of their operations. Young landowners represent between 6% and 13% of each landowner group: Asian Americans make up 1% of all younger landowners, and younger landowners comprise 12% of all Asian American landowners. American Indians or Indigenous make up 2% of all younger landowners, and they comprise 13% of all American Indian or Indigenous landowners. Black or African Americans make up 1% of all younger landowners, and they comprise 6% of all Black or African American landowners. Hispanics make up 4% of all younger landowners, and they comprise 13% of all Hispanic landowners. Non-Hispanic Whites make up 92% of all young landowners, and they comprise 12% of all non-Hispanic White landowners. Younger groups are not entering farming as their primary occupation. It was important to mention age in this section, as it relates to experience, and younger landowners are generally by default less experienced. Since age was previously described in detail, beginner landowners will be described further here, as they subsume younger landowners as well.

At a regional level, experience does not vary significantly: Larger operations and operations in higher economic class categories tend to be stewarded by more experienced landowners, those who are older than 35 years and who have greater than 10 years of experience (figures 6.29–6.31); thus, federal programs targeting younger and less experienced landowners were created to facilitate technical assistance.

What does this mean?

Beginner landowners are typically middle-aged and older, perhaps beginning a new career in farming or supplementing their current employment. They may not necessarily have a farming background and may migrate to rural areas from urban communities; however, a third currently reside off-farm. Program development associated with off-farm residence and farm management may be beneficial to attracting these two groups.

When developing future programming, determining commodities that are both profitable and appealing to less experienced and younger landowners may be beneficial overall, as they, in part, will replace aging landowners retiring soon. Commodity-based and stewardship-related training, focusing on those that yield indirect or direct profit, along with those associated with supplemental enterprises and incentive-driven programming, may make land ownership and farming more appealing to less experienced and younger landowners. Programming associated with ecosystem benefits derived from their land-based operations would benefit landowners and nearby communities.

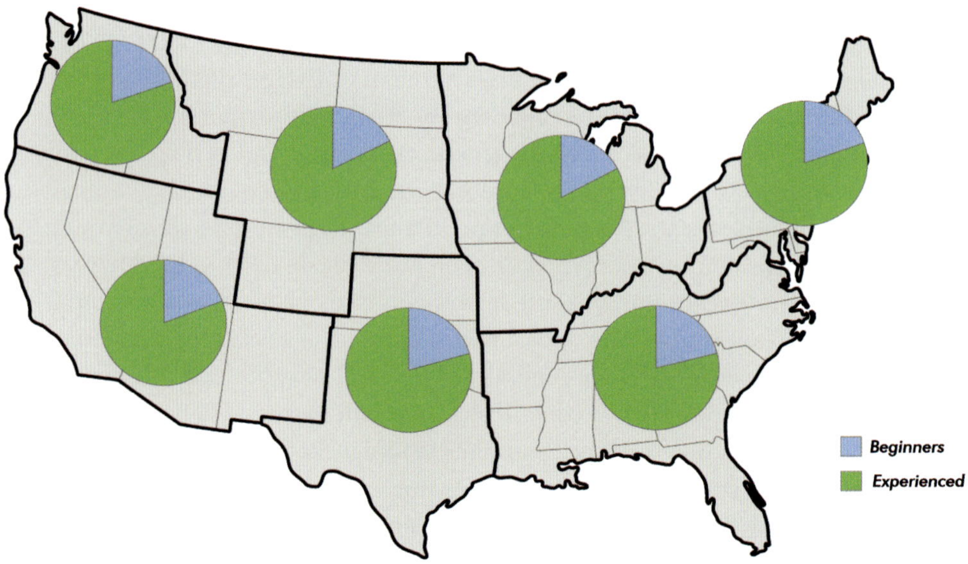

FIGURE 6.29. Landowners by operational experience (years) and region, 2017.

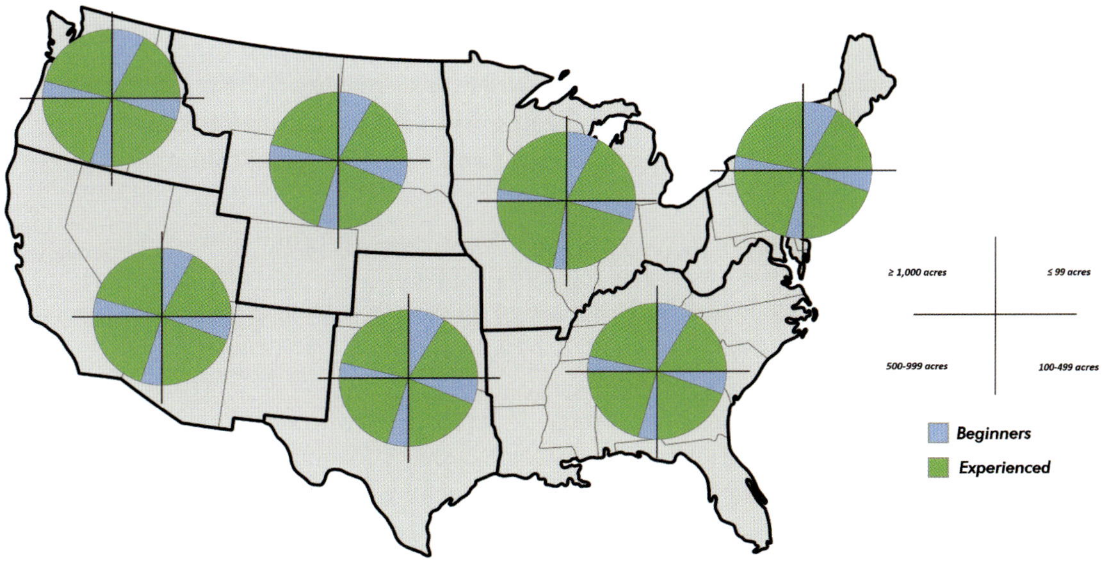

FIGURE 6.30. Landowners by operational experience (years), operation size, and region, 2017.

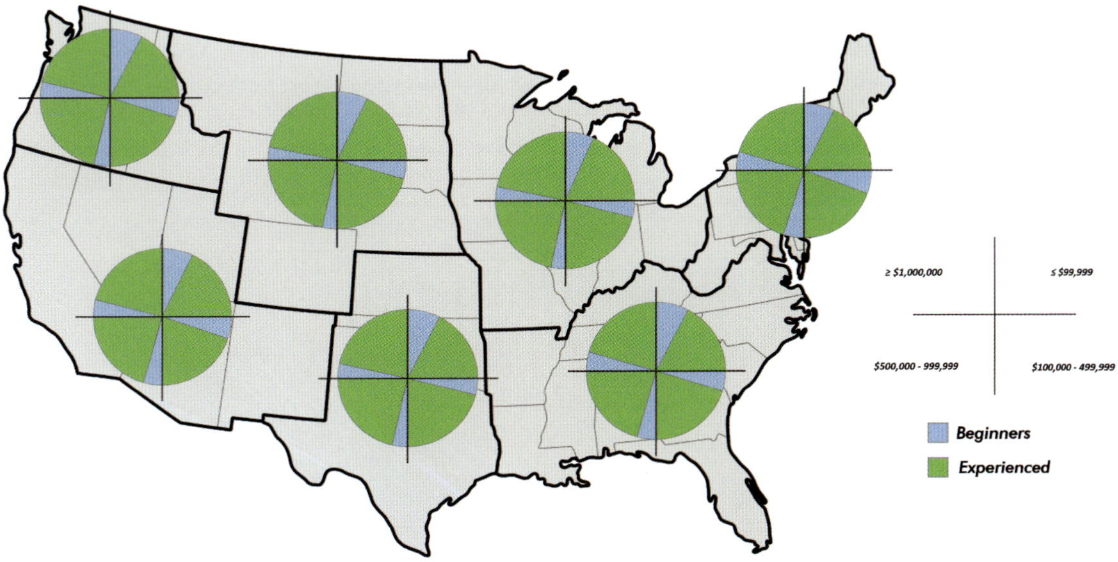

FIGURE 6.31. Landowners by operational experience (years), economic class, and region, 2017.

Residence

Historically, landowners primarily lived on their farms and managed their operations on-site. Now, landowners who reside off-farm (26%) also manage successful agricultural enterprises. There is a perception that off-farm residence (absenteeism) is associated with "differ[ent] values and land management behavior from those who live fulltime on their land" (Sorice et al. 2018, 160). They are considered to be less invested in the land, engaging in less than adequate land management, or possessing limited commodity management skills. Although off-farm residence is influenced by many factors, developing active, targeted engagement strategies that meet absentee operational needs is recommended (Snyder et al. 2020).

Regionally, the largest percentage of landowners who reside off-farm (25%–30%) are found in the Southwest, Southern Plains, and Southeast regions, while in the Pacific Northwest, Northern Plains, Midwest, and Northeast, 17%–25% of landowners reside off-farm (figures 6.32–6.34). The Pacific Northwest had the least percentage of landowners residing off-farm, followed by the Northeast in the 99 acres or less farm size category (i.e., 14%–18%). The southern half of the United States had the largest percentage of off-farm residence. The Southwest had the largest percentage of landowners residing off-farm in operations between 100 and 999 acres (40%–42%), except in the 1,000 acres or more category, where the Southeast had the largest percentage (39%). The Southwest also had the largest percentage of

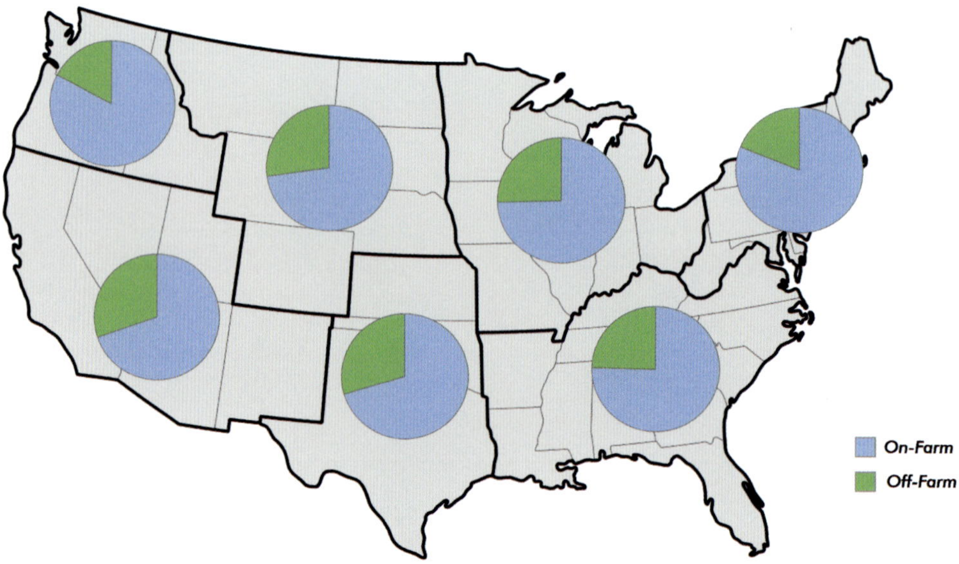

FIGURE 6.32. Landowners by residence and region, 2017.

FIGURE 6.33. Landowners by residence, operation size, and region, 2017.

FIGURE 6.34. Landowners by residence, economic class, and region, 2017.

landowners who resided off-farm in economic class categories $100,000 and greater (42%–56%), while the Pacific Northwest and the Northeast had lowest percentage of landowners who resided off farm in the $99,999 or less economic class category (15%–18%).

Most off-farm residents are non-Hispanic White: 0.3% are American Indian (20% of all American Indian or Indigenous landowners reside off-farm), 0.2% are Asian American (41% of all Asian American landowners reside off-farm), 0.5% are Black or African American (39% of all African American or Black landowners reside off-farm), 1% are Hispanic (35% of all Hispanic landowners reside off-farm), and 24% are non-Hispanic White (25% of all non-Hispanic White landowners reside off-farm). New and beginning landowners and young landowners, 34% and 33%, respectively, reside off-farm, along with 23% of landowners with military experience. Female (21%) and male (28%) landowners also reside off-farm and successfully manage their land-based agricultural enterprises.

Off-farm residence may be attributed to many factors, such as secondary employment, or it simply may be commodity driven (i.e., intensive vs. extensive agricultural production). Off-farm residents also are uniformly present in all operation size and economic class categories; thus, it is important to point out that off-farm resident landowners across the United States also are successful land and commodity managers.

What does this mean?

Most landowners live on their operations (74%) with slight differences regionally. Programming efforts should consider developing specific resources for absentee landowners, based on their operation's management objectives. Engaging off-farm residents in conservation programming is suggested and also beneficial for on-farm residents. Improved precision agricultural technologies may benefit off-farm residents; however, assessing whether and which commodities influence off-farm residence is a first step to creating more effective technologies, along with determining their need, usefulness, applicability, and ease of adoption. Effective, service-oriented communication strategies would benefit both landowner groups.

Primary occupation

Across the United States, 58% of all landowners are employed *off-farm*, meaning, their land-based operations are not their primary occupation or their primary source of household income (female, 66%; male, 54%; new and beginning, 72%; young, 65%; military, 67%; Asian American, 50%; American Indian or Indigenous, 51%; Black or African American, 56%;

Hispanic, 60%; non-Hispanic White, 59%). The Northern Plains and the Southwest have the highest percentage of landowners listing farming as their primary occupation (50%). Farming as a primary occupation increases as farm size and economic class category increase (figures 6.35–6.37). Little information is available describing the reasons for seeking off-farm employment, whether a primary occupation other than farming is associated with land inheritance (i.e., inheriting land after starting an off-farm career), is commodity-driven (i.e., agricultural commodity does not yield a sustainable income), or is necessary to keep the farming operation afloat. A shift toward owning land for reasons other than farming may have occurred, such as wildlife- and natural resource–related recreation or maintaining land for future generations (Lopez et al. 2023). Exploring and understanding these decisions may be beneficial to developing working solutions and programming for part-time farming families.

What does this mean?

Farming as a primary occupation is on the decline (USDA 2023). This may be associated with a variety of factors, from the automation of larger-scale farming practices to the need to supplement farming-related income; however, it does not explain the widespread adoption of farming as a nonprimary employment option for landowners. It would be relevant to determine farming or landowning incentives already in place that are helping landowners maintain their operations. This is promising, as perhaps the incentives associated with ecosystem benefits, among many others, including locally specific benefits, may be appealing to landowners for their own benefit and that of nearby communities.

Locally specific benefits are relevant to program development, as this demographic review is at a national and regional level, and what may benefit one area may not benefit another. For example, here we determine that farming as a primary occupation is not increasing in interest for future generations; otherwise, the number of younger farmers would be much greater. We have also seen that farming as a profession is not as profitable; otherwise, the income derived from farming would be much greater. Since these are data-driven facts, it seems that incentives that make farming both more profitable and more enticing to younger generations are necessary. Benefits that the land automatically provides, without much effort, such as nature benefits, are enticing as incentives to farming because they are additive; that is, they are in addition to income derived from produced agricultural commodities.

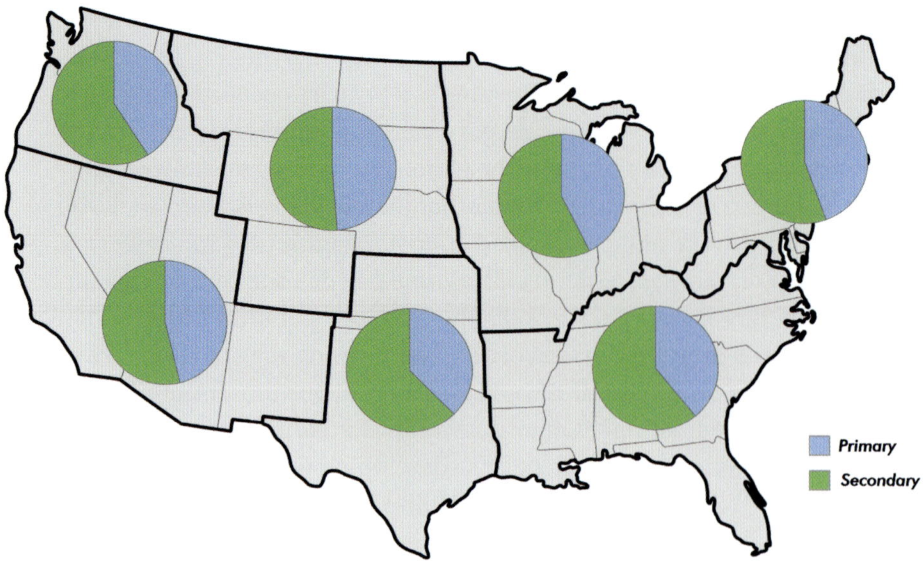

FIGURE 6.35. Farming described as a landowner's primary occupation or as a secondary occupation (farming is not primary occupation) and region, 2017.

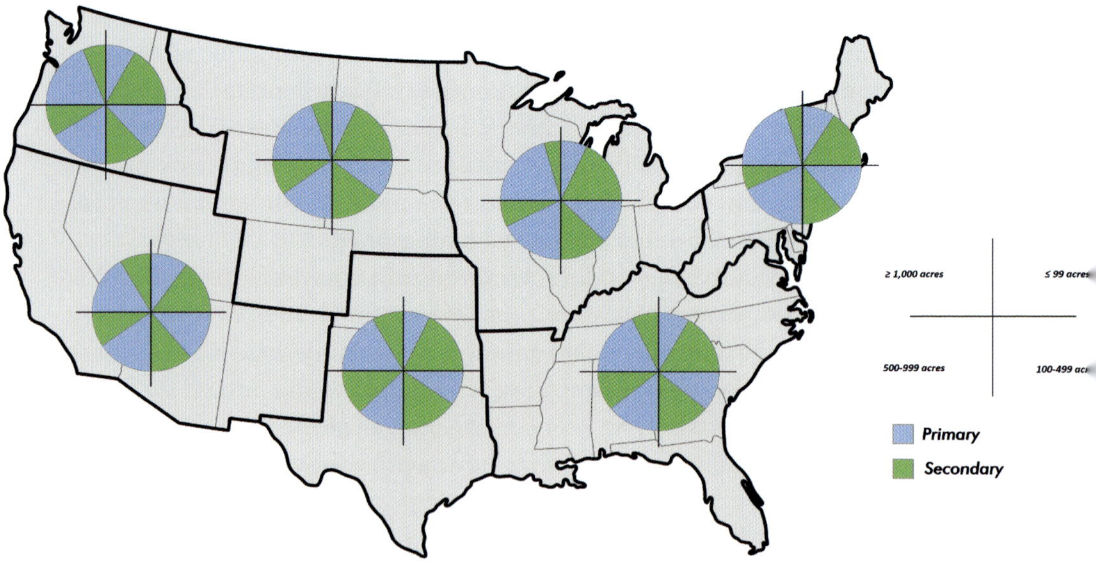

FIGURE 6.36. Landowners by primary and secondary occupation, farm size, and region, 2017.

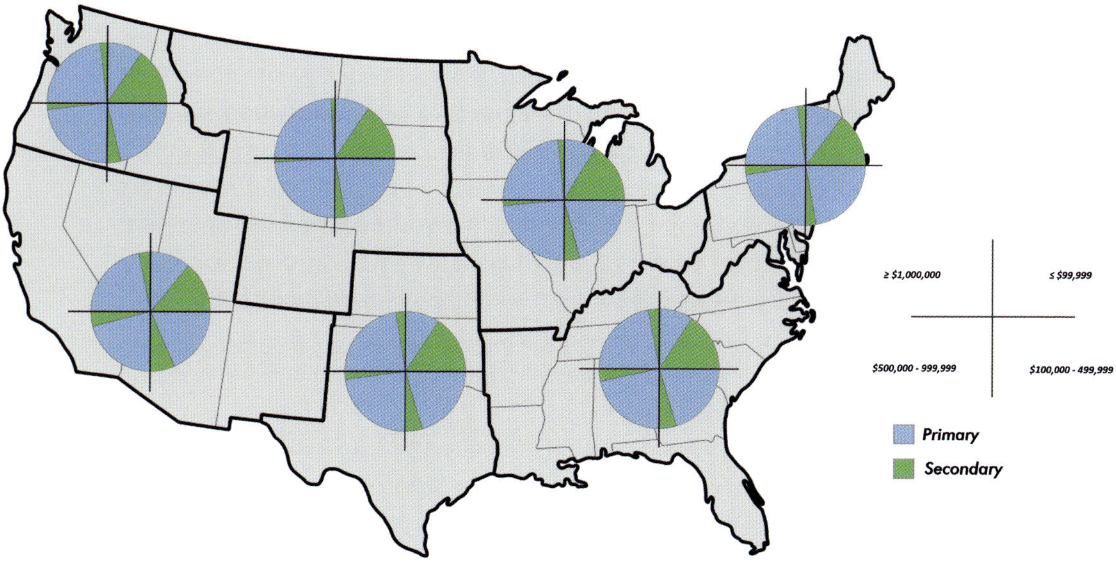

FIGURE 6.37. Landowners by primary and secondary occupation, economic class, and region, 2017.

Conclusion

Rural working landowners play a crucial role in providing many community benefits, including food, fiber, outdoor recreation, and ecosystem benefits. Natural resource professionals and collaborative organizations play a crucial role in ensuring that landowners and communities mutually assist and benefit one another. Targeted programs and informational resource development may both provide technical assistance and be a reminder that they live in a shared natural resource landscape that must be carefully managed. Effective programming incorporates the demographic composition of communities and landowners at a local, regional, and national level, as well as the interplay of outside forces and their influence on communities, landowners, markets, agricultural commodities, and natural resources.

Landowners control and influence the nature benefits derived from their lands that touch nearby communities. Understanding landowner demographic characteristics helps natural resource professionals create effective programming that is relevant, useful, and focused on the intended audiences and their needs. As examples, apart from technical assistance programming, some USDA programs provide information and financial incentives to agricultural and natural resource professionals and to all landowners, including those nearing or at retirement age, and to young, new and beginning, military, female, and socially disadvantaged landowners, who are more likely to own smaller properties and to derive less household income from their land. Creating strategies that help landowners successfully manage their agricultural and natural resource enterprises; derive more income from a variety of sources, such as conservation incentives and new markets; and build collaborative networks to improve operation and land stewardship practices benefits landowners and nearby communities. Building strong rapport, based on access to networks and active listening skills, translates landowner and community needs into efficient programming and lasting collaborative bonds into more resilient natural resource–based communities.

"The quest for simple and reliable indicators to help explain land management behavior is worthwhile and important for understanding how changes in land ownership demographics may affect land management decisions and the corresponding provision of ecosystem services that are fundamentally important for rangeland health and human wellbeing" (Sorice et al. 2018, 162).

CHAPTER 7

Values and Attitudes

JENNY C. SANDERS AND ROEL R. LOPEZ

This chapter introduces the reader to a national land stewardship typology framework that illuminates the dominant value systems found in the land steward community. Insight into the values that guide the motivations and decisions of land stewards can lead to deeper relationships that result in synergies between practitioners and land stewards seeking partnerships to work for the greater good of land management and conservation.

Introduction

The previous chapter introduced national and regional trends in current landowner demographics. In this chapter, we discuss the evolution of the *value systems* and *attitudes* of private land stewards and describe the role of a landowner typology framework in understanding the connection of landowners to their lands, including what motivates their decisions and interests. Private landowners are not homogeneous but instead have very diverse perspectives, complex backgrounds and experiences, and differing goals and objectives for their properties.

Values are basic and fundamental beliefs that guide or motivate attitudes or actions. The value systems that guide private landowners are influenced by many factors, including geographic, demographic, cultural, sociopolitical, and economic characteristics. Previous researchers have attempted to develop and describe landowner typologies (i.e., classification system based on types or categories) based on some of these drivers. Here we hope to accomplish three objectives: (1) provide a historical review of landowner stereotypes (1800s–present), (2) present a landowner typology model framework based on a landowner's relationship to the land, and (3) illustrate how other landowner typology studies relate to the proposed model framework presented.

Historical View of Landowners

America's frontier was settled by unique individuals who immigrated to various places and were guided by a shared vision to "conquer" the wilderness and to feed and clothe the society that would follow them (Peterson 1990). This was considered a noble calling and framed the early views of many private landowners. This historical view of landowners was rooted in a "spiritual" call to subdue the land, with a foundation built on the values of hard work, independence, and family (Valencius 1969).

Private landowners have evolved in their cultural view of self over time from what a previous study describes as America's "frontier heroes" who originally settled our continent, to the ranchers and farmers of today (Peterson 1986). From Leopold's suggestion that a "land ethic changes the role of humans from conqueror of the land-community to plain member and citizen of it" (1949, 205), we see the basis for the apparent paradigm shift in the judgment-guiding value sets of contemporary landowners. As a predominantly agrarian society during the "frontier heroes" era, producers and operators in their haste to produce short-term economic gains were often unaware of the long-term impacts aggressive and detrimental farming practices had on the soil and land held in their care (Peterson 1986). The same was true of stockmen, where number of cattle, sheep, and goats far exceeded the grazing capacity of the land during this period. The end result was devastatingly clear during the Dust Bowl years of the 1930s, where drought coupled with poor agricultural and land-use practices created compounding natural resource disasters for both humans and wildlife (Peterson 1986, 1991). To make matters worse, government agency managers advanced the notion that climate was solely to blame for damages suffered on these rural working lands. As would logically follow, farmers and the agencies that provided farm programs were released from any sense of responsibility for damage to the land and would often adopt conservation measures only as long as their short-term goals were met. Consequently, practices were abandoned when short-term economic results were no longer evident (Peterson 1991).

It was not until the mid-1940s that the concept of stewardship began to become more prominent in American agriculture. The American Country Life Association (ACLA), for example, formed under the Theodore Roosevelt administration, was largely responsible for the promotion of the idea of stewardship to rural America with its 1946 "Statement of Principles," which stated that "ownership of land does not give an absolute right to use or abuse, nor is it devoid of social responsibilities. It is in fact a stewardship"

(Wunderlich 2004, 7). Similarly, many rural churches that historically described stewardship in terms of one's body and monetary possessions began to expand the idea to include all of nature and God's creation (Peterson 1991; Wunderlich 2004). Stewardship as a formal agricultural concept began to evolve from this point forward.

Many changes have taken place since the expression of ideals and motivations of Dust Bowl–era landowners. Regrettably, few documented or formal studies described contemporary landowner perceptions of self and their responsibility to the land until the early 1990s. Consequently, today's landowners are often stereotyped as being focused primarily on short-term economic profitability and having little or no concern for ecological sustainability. This is an inaccurate description of many of today's landowners. For natural resource practitioners, administrators, and policymakers, failure to understand the motivations of today's private landowners can hamper the formation of potential agriculture, wildlife, and natural resource conservation partnerships between state or federal government entities and other landowner advocacy groups. Understanding the motivations and values of private landowners serves as an important first step toward productive partnerships that result in sustainable agriculture and natural resource conservation on private lands.

A Landowner Typology: Relationship to Land

What are some characteristics and values of the modern landowner? Some have described landowners as being driven by common sense, independence, and human-land connection and having much respect for generational and experiential knowledge (Peterson 1990; Peterson and Horton 1995). Other studies have noted that many landowners express a distinct agricultural land ethic, describing, for example, "ranching as a lifestyle formed by making a living, being on the land, and taking care of the land" (Williams 2000).

Many landowners also emphasize long-term goals for their property and the pride they feel in their ability to pass their land along to future generations. Contemporary landowners often express disdain for the image of the frontier hero, equating the frontiersman image to one of poor management practices and exploitation of natural resources (Peterson and Horton 1995). Numerous studies nationally have reported decreasing numbers of landowners who list economics as their primary motivation for land ownership and management and an increasing number who are more focused on aesthetics and quality-of-life aspects of owning land. Although promising from a conservation standpoint, natural resource practitioners must be open-minded and careful listeners as they make determinations about a landowner's motives relative to conservation, economics, and how best to engage in future discussions to better understand an individual landowner's decision matrix for land management.

As a practitioner, it is important to understand that landowners have many similarities but are not entirely a homogeneous group. Factors such as tenure on the land, location and size of property, financial resources, and family heritage can lead to different motivations and behaviors related to land management. We propose the following framework that uses landowners' relationship to the land as the primary descriptor of their stewardship motivations and management decisions: (1) *Born to the Land*, (2) *Reborn to the Land*, (3) *Ag Business*, and (4) *Asset Owners*. As you become familiar with these typologies, you will likely find them informing all aspects of how you relate to land stewards. Think of this framework as a cheat sheet for success in your job. If you are able to quickly identify which of these profiles an individual or group of landowners falls into, you will immediately know how best to appeal to their values and, therefore, develop a productive relationship.

This typology framework was originally developed in Texas (Sanders 2005), but we see opportunity for nationwide application based on a comprehensive review of similar studies conducted throughout the United States (table 7.1), which we discuss later in the chapter.

Broad Landowner Groups

Born to the Land

> *All I hope is that my grandkids have a high respect for the land. . . . The good Lord said you are a steward of the land, and I believe you have a responsibility to do what's right for it.*
>
> **—Texas Born to the Land Owner**

This group exhibits a very strong connection and/or sentimental value to their land (figure 7.1). They often speak fondly of the generations before them who have farmed or ranched the family property. Agricultural production is often a primary source of income, and although many recognize the potential benefit of emerging or nontraditional enterprises, they generally focus their efforts on a more traditional way of life. These landowners exhibit a strong sense of stewardship or responsibility for the land, many times expressing stewardship as a spiritual calling.

Landowners within the Born to the Land category generally express caution and wariness about what they consider to be "government programs." They wish that farming and ranching were not subsidized; however, most of them have participated in some form of subsidy, the most common being drought relief or other disaster-related payments. This disparity between values and needs of landowners is a common thread throughout the Born to the Land group. They tend to strongly desire independence from the government but many times do not have the financial resources needed to back up that desire. They, therefore, prefer short-term, locally implemented programs. Additionally, many in this group have a strong desire to do what is best for their property, so programs that they interpret as being designed to empower them to do those things are more attractive than ones that seem to promote specific government agency objectives. In terms of educational opportunities, Born to the Land owners tend to appreciate field demonstrations where they can learn from other landowners and practitioners by viewing actual projects.

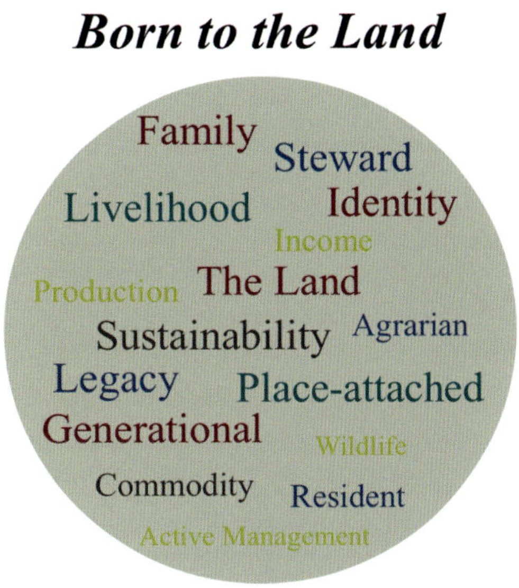

FIGURE 7.1. Common defining characteristics and descriptors for the Born to the Land profile group.

Reborn to the Land

I've been around the world a little bit, and having a piece of land is such a gift. . . . I feel that I want to protect it, and I want to make it healthy and be able to pass it on to our kids.

—Texas Reborn to the Land Owner

This group is also characterized by a strong connection to the land (figure 7.2). As the name suggests, individuals in this group have some kind of historical connection to the land or agriculture and have returned to "their roots" either in retirement or as a family retreat. Their focus is on the

FIGURE 7.2. Common defining characteristics and descriptors for the Reborn to the Land profile group.

aesthetic and recreational value of the land, and they feel strongly that they have a responsibility to "take care of it."

Agricultural or wildlife enterprises in place on these properties are generally more for personal satisfaction than economic gain. In most cases, these landowners are not worried about generating a profit, yet some still operate on a limited budget. A small property landowner emphasizes this point: "Yeah, that's really why I do it—it's for enjoyment. As a matter of fact, there's probably no way I would ever turn a profit if you consider all the money we're putting into it." Another gentleman expressed similar feelings when he said, "Unfortunately, it's not the money that inspires me to keep the land. . . . It's other things."

One characteristic that sets the Reborn to the Land group apart from others is their acknowledged need for information about government financial and technical assistance programs to support their management and conservation activities. These landowners are interested in being "taught" about management but generally do not know where to go for information. For this reason, many of them actively seek educational opportunities and knowledge that could help them manage their properties. They have a passion for "doing what's right" on their property, and they are willing to listen to anyone who can give them sound information.

Reborn to the Land owners are the most likely to respond positively to the widest range of government programs, yet they are the least likely to have participated in programs in the past. Given that landowners in this group are commonly focused on the recreational and aesthetic aspects of their property, and therefore put less emphasis on the profit margin their operation generates, they generally have the least familiarity with natural resource agencies and programs.

This group is also likely to obtain a large degree of management information from incidental conversations with neighbors and agriculture professionals (e.g., feed store, chemical or seed company employees, equipment operators). They are also more likely to obtain information from the popular media (i.e., magazines, newspapers, the internet, and even television and radio) than the other sources. While many in this group are willing to participate in subsidy and cost-share programs, they are most interested in technical assistance and educational programs. Any opportunity to increase their knowledge and skills about management of their property is commonly more appealing to this group. Finally, Reborn to the Land owners are generally receptive to programs that might create notoriety for them in their conservation journey. They are many times motivated by restoration of historical ecosystems and creating a natural legacy for future generations.

Ag Business

> *We're a combination wildlife operation: dove, exotics, trophy deer, and a bow operation. Guys from all the states come in here. . . . We have a cook—it's a full-blown operation!*
>
> **—Texas Ag Business Owner**

> *I love it. It makes you feel good. It gives you something to get up every morning and do. . . . It gives you satisfaction and it's a challenge to try to make a profit on it.*
>
> **—Texas Ag Business Owner**

These landowners do not articulate a strong connection to the land and as a group tend to be much more open to the thought of selling their land than other profile groups (figure 7.3). A particular landowner in the Ag Business profile articulated this attitude when he said, "There will always be a Wilson Ranch. But, I don't know if it will always be right here." This group of landowners is more connected to the "business" side of land ownership than to the particular piece(s) of geography on which the operation is located.

Agricultural production and wildlife enterprises are a source of pride for these individuals but are rarely a significant source of their income. Ag Business landowners generally come from successful business backgrounds and are presently focused on creating a profitable agricultural business.

Relative to incentive or subsidy programs, they tend to have similar values as those in the Born to the Land group; however, they are set apart by the financial resources they have within their grasp. This fact is articulated well by an absentee landowner whose day job as an attorney

FIGURE 7.3. Common defining characteristics and descriptors for the Ag Business profile group.

FIGURE 7.4. Common defining characteristics and descriptors for the Asset Owner profile group.

supplies the financial resources he utilizes to improve his ranch and cattle operation. He said, "People that are financially independent, I would think, would have less use for government programs." This same gentleman went on to say that he appreciated getting management information from the government, but he just really was not interested in the hassle involved in participation in government programs. "There are too many layers of bureaucrats, so I'll just get the information and do it myself, the way I want to do it. When I want to do it, I mean, I'll listen to them, but I just get really frustrated between levels of control." Ag Business landowners will be most likely impacted by assistance and personal attention by those they see as experts in a field in which they are interested (e.g., private consultants, university professors, Extension specialists, or other noted experts).

Asset Owners

This fourth group of landowners has very little connection to the land at all (figure 7.4). Many times, they inherited their properties, and while they see investment value in the property, or perhaps receive mineral revenue from it, they tend to have very little interest in any kind of active management. This type of landowner may be the toughest to reach with a conservation or management message.

Summary: Overlapping Values

Landowners come from various backgrounds, have differing levels of income and experience, and have many different goals and reasons for ownership of land. Consequently, the historically monolithic stereotype of landowners is no longer an effective framework on which to build natural resource policy or guide interactions with landowners. The increasingly complex climate of land ownership necessitates strong reevaluation of programs and services offered to landowners throughout the nation and the way those programs and services are marketed and promoted to different groups of landowners.

Born to the Land, *Reborn to the Land*, *Ag Business*, and *Asset* landowners each have distinct values, goals for their property, and ideals that guide their land management decisions but also have some overlapping values that should be noted (figure 7.5). Many times an individual's self-perceived identity as a farmer or rancher can be used as an important means of

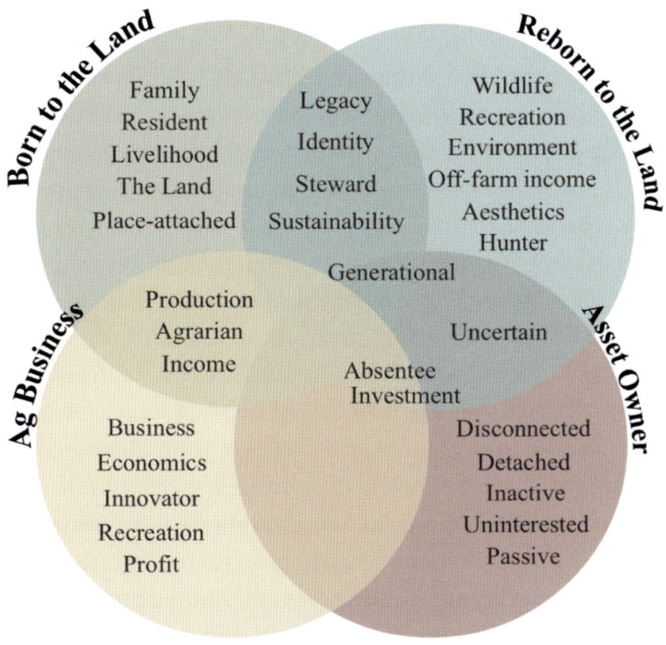

FIGURE 7.5. Diagram depicting unique and overlapping characteristics and themes relative to each of the four land stewardship typologies.

tapping into their value systems and ultimately their motivations for land management. Here is a summary of these four major groups:

Born to the Land landowners feel strongly that their identity as a farmer or rancher carries with it a responsibility to uphold the traditions and heritage of agriculture and the rural lifestyle that they cherish. They also possess a strong desire to steward the land in their care to the best of their ability, ensuring its availability for utilization and enjoyment by future generations. Their major objective for their land is to produce and sustain a living through a rural, agricultural lifestyle.

Reborn to the Land landowners envision themselves as "saviors" of the rural, farming, or ranching lifestyles and stewards of the land in their care. Their major goal is to produce a recreational lifestyle that simulates farming or ranching and a romantic stewardship experience on their land.

Ag Business landowners also cherish agriculture and their identity as farmer or rancher. These landowners feel that their non-ag background offers something new to the agricultural community and that their business emphasis is what is lacking in many agricultural endeavors. Their major objective is to be identified as a landowner and to prove that one can make a profit from a progressive entrepreneurial enterprise.

Asset Owners are the most difficult to profile and reach, but practitioners must not discount the value of land held by Asset Owners, especially in a climate of emerging economic opportunities and rapid generational transfer. Each piece of land holds unique value, and each landowner has a unique touchpoint for engagement in stewardship and conservation. While this group is far from the "low-hanging fruit" sought out by most conservation programs, they will likely be interested in programs or opportunities that could add value to their land asset without creating an undue financial or effort burden on them. Emerging ecosystem marketplaces, mitigation programs, and renewable energy development are good examples of opportunities for possible engagement of Asset Owners.

Landowner Typology: A National Model

Optimistic that this framework successfully deployed in Texas would be applicable nationwide, we conducted a qualitative literature review and assessment of common terms described in various research efforts that evaluated regional landowner values across the United States. The goal of this exercise was to validate our framework as a standard that could be applied nationwide to anticipate likely landowner response to various approaches and programs in order to accelerate cooperative conservation efforts.

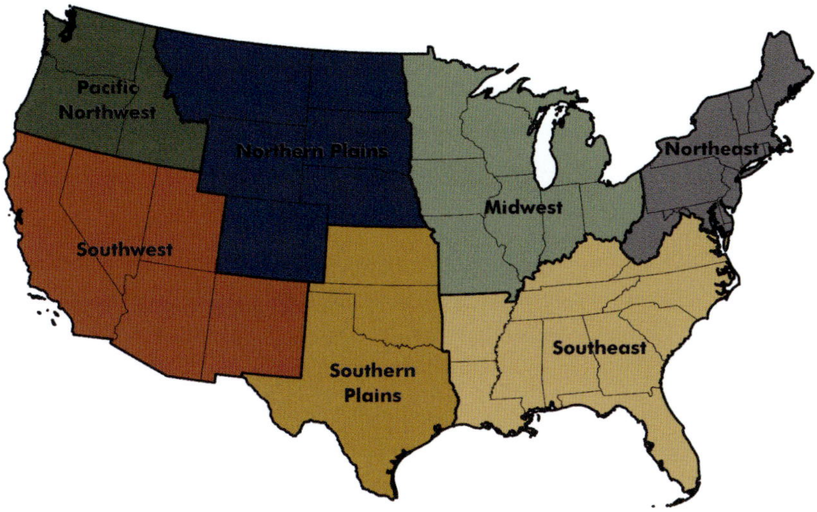

FIGURE 7.6. US regions used in characterizing private lands.

Table 7.1 represents the results of this exercise, in which we list landowner typology studies from each of the regions identified in chapter 4 (figure 7.6) and color-coded them to match figure 7.5.

While most of these studies were narrowly focused on particular programs or geographic locales, analysis of their typology descriptors shows strong overlap with our typology based on relationship to the land and could be used universally.

For example, Sweikert and Gigliotti (2019) utilized a Land Use Value Scale (LUVS) to understand the conservation decisions of agricultural producers in the Northern Plains and Midwest. Their work identified four LUV types rooted in a perception that humans or nature is first in their decision-making process. Their categories align with our framework as follows:

1. Interconnected, legacy owners (Born to the Land): Value both the environment and agriculture. View land as an important part of their personal and family identity.
2. Nature first, interconnected (Reborn to the Land): Value both the environment and agriculture. Strongly motivated by the environment and wildlife.
3. Humans first (Ag Business): Focused almost exclusively on crop production and financial viability.
4. Disconnected (Asset Owner): While the authors did not elaborate on the characteristics of a disconnected owner, the name implies overlap with what we call an Asset Owner—someone with a weak underlying ethos for their land ownership who sees land as merely another asset in their investment portfolio.

Similarly, Balukas et al. (2019) classified landowners in urbanizing forested landscapes in the Northeast to better understand conservation opportunities. Their analysis revealed the importance of "place attachment" in informing the attitudes and motivations of landowners. They classified landowners in three categories that align with our framework as follows:

1. Place-attached managers (Born to the Land): Strongly attached to the land and consider forest production important.
2. Place-attached stewards (Reborn to the Land): Indicate a sense of responsibility for stewardship but generally do not own land for forest production.

   Wildlife stewards (Reborn to the Land): Do not have strong attachment to place but feel a sense of responsibility to manage the plants and animals on their land.

TABLE 7.1. Summary of landowner typologies conducted in the United States compared to our Land Stewardship Typology Framework, 1991–2019

BORN TO THE LAND	REBORN TO THE LAND
Interconnected: Value both the environment and agriculture. **Legacy owners:** View land as an important part of their personal and family identity.	**Interconnected:** Value both the environment and agriculture. **Nature first:** Strongly motivated by the environment and wildlife.
Legacy owners: View land as an important part of one's personal and family identity.	**Legacy owners:** View land as an important part of one's personal and family identity.
Multi-objective: Value aesthetics, investments, and nontimber goals from their forestland. **Timber-objective:** Emphasize investments, recreation, and aesthetic goals from their forestland.	**Amenity:** Value aesthetics and recreational goals from their forestland. **Multi-objective:** Value aesthetics, investments, and nontimber goals from their forestland.
Agrarian/Economic: General low interest in wildlife management and see land as a source of income. **Agrarian/Steward:** Live on their property and are focused on agricultural production.	**Absentee hunters:** Mostly use their land for hunting while residing elsewhere. **Residential landowners:** Live on the property and mostly use it for residence and recreation, not income. **Next-generation landowners:** Most likely to have inherited their land, have the highest off-farm income, and have the most interest in wildlife management but have implemented few wildlife-friendly practices. **Wildlife managers:** Value owning land to hunt and fish and also like to help wildlife.

3. Passive landowners (Asset Owners): Do not have a strong attachment to place, an articulated stewardship ethic, nor do they own their land for production purposes.

Balukas et al. do not identify a category that aligns with our Ag Business typology; however, they do state that "some landowners hold stewardship values while not having a strong sense of place attachment to a particular parcel, making them more likely to sell their land than other stewards" (756). This statement is very much in line with our description of an Ag Business owner, so we feel that the typologies align in this case as well.

Table 7.1 offers many other examples of localized studies that, when compared to our land stewardship typology framework, provide validation that it is a model that has universal, nationwide utility for using a landowner's connection to the land as a predictor for the owner's motivations, attitudes, and actions related to land stewardship.

AG BUSINESS	ASSET OWNER	CITATION
Humans first: Focused on crop production and financial gains.	**Disconnected:** Least involved in agriculture and reside in urban areas.	Sweikert and Gigliotti 2019 (Northern Plains and Midwest)
Detached owners: Generally disengaged from their land and from their role as forest owners.	**Detached owners:** Generally disengaged from their land and from their role as forest owners.	Richter 2005 (Midwest)
Timber-objective: Emphasize investments, recreation, and aesthetic goals from their forestland.		Khanal et al. 2017 (Southeast)
Agrarian/Economic: General low interest in wildlife management and see land as a source of income.	**Uncertain landowners:** Have few strong opinions on land and not interested in wildlife.	Schweiss and Dwyer 2008 (Midwest)

TABLE 7.1, CONTINUED Summary of landowner typologies conducted in the United States compared to our Land Stewardship Typology Framework, 1991–2019

BORN TO THE LAND	REBORN TO THE LAND
Working the land: Multi-objective owners who appreciate amenity valu es of the land but also rely on production income.	**Woodland retreat:** Most likely to live on their land, which was usually a smaller parcel, and considered the amenity values most important.
Multi-objective: Strongly motivated by consumptive and nonconsumptive uses equally. **Timber:** Interested only in timber management and land investment.	**Non-timber:** Place more importance on aesthetic values, biodiversity, and recreation. **Multi-objective:** Strongly motivated by consumptive and nonconsumptive uses equally.
Commodity managers: In addition to timber production and grazing, also frequently identify family legacy and privacy as very important goal.	**Amenity managers:** Most frequently identified residence, privacy, aesthetics, and habitat as very important goals. **Recreational managers:** Absentee landowners, but held amenity goals as very important (e.g., privacy, aesthetics and family legacy).
	Professionals: Living simply is important motivator to land ownership. **Preservationists:** Value a country lifestyle. **Young families:** Motivated by the lifestyle provided from their land for the benefit of their family. **Farmers:** Value farming and personal finance, but lifestyle motivations (living simply, naturalism, ruralism) are most important to them. **Forest planners:** Similar motivations to farmers, but just do not have time to implement practices yet.
Poor rural residents: Inherited small parcels. Likely to sell timber or do other income-generating practices. Unlikely to participate in other active forest management.	**Timber managers:** Most active managers, most likely to sell timber, and plan to pass land on to future generations. Own land for investment, recreation, and hunting purposes. **Resident conservationists:** Least likely to sell timber. Own land for environmental reasons, to preserve wildlife habitat, and participate in a rural lifestyle. **Affluent weekenders:** Absentee owners. Own property for weekend home and investment. Not interested in timber income. Motivated by wildlife conservation and nonconsumptive recreation.
Timber agriculturists: Dedicated to sustained timber production. Believe responsible timber management also benefits wildlife and provides other amenity benefits. Business-oriented.	**Timber conservationists:** Less business oriented than timber agriculturalists but also dedicated to timber management, especially for the benefit of wildlife, and aesthetic benefits. Feel responsible for maintaining the forest system. **Forest environmentalists:** Regard the forest as a reservoir for a host of benefits: beauty, privacy, wildlife, etc. Not interested in timber production. Ecological stability is important.

AG BUSINESS	ASSET OWNER	CITATION
Supplemental income: Most active management, most likely to harvest trees and participate in various programs.	**Ready to sell:** Older, absentee owners not actively managing property.	Butler et al. 2007 (USA)
Timber: Interested only in timber management and land investment.	**Timber:** Interested only in timber management and land investment.	Majumdar et al. 2008 (Southeast)
	Passive managers: Mostly absentee owners that did not identify any of the listed goals as important for their ownership of the property.	Fischer 2012 (Pacific Northwest)
	Absentee investors: Economics most important motivation.	Kendra and Hull 2005 (Southeast)
		Kluender and Walkingstick 2000 (Southeast)
Range pragmatists: Business and production oriented. Appreciate wildlife values but will convert to rangeland if economically advantageous.		Kurtz and Lewis 1981 (Midwest)

TABLE 7.1, CONTINUED Summary of landowner typologies conducted in the United States compared to our Land Stewardship Typology Framework, 1991–2019

BORN TO THE LAND	REBORN TO THE LAND
Sustainable harvesters: Rely on long-term balanced land use. Sustainability is important, as they hope to pass their land along to future generations. **Back 40ers:** Forest resources are important, but focus is on shorter-term agricultural production.	**Recreationalists:** Value their forestland as a place for personal use and do not see their land as a source of income nor an economic burden.

Source: Sanders 2005

Conclusion

Agencies and organizations with a mission to conserve wildlife and natural resources through private land management will be well served to take seriously the implications of this framework. Born to the Land owners, as a group, are most likely to respond positively to programs that empower and enable them to maintain their lifestyle and stewardship responsibility. Reborn to the Land owners, while they will likely be the most receptive to the widest range of programs, are largely unaware of the programs and services available to them from agencies and organizations. Additionally, due to their admitted naivete regarding agriculture and land management in general, it is of utmost importance that they be reached with sound management principles early in their land management career. Outreach is important with the Reborn to the Land group. Ag Business landowners, while largely unwilling to participate in government programs, will be most likely impacted by short-term programs, personal attention by recognized individuals in a particular field that interests them, and/or private technical service providers.

While not all landowners express an identifiable land ethic, many do, and the strength of it creates a unique opportunity for agencies and organizations to use the knowledge and understanding of landowners' values to "sell" particular programs through strategic packaging and/or marketing. Further, it is important to remember that time, cultural maturity, and ethical nurturing are key factors as landowners develop a land ethic. Therefore, there is opportunity to reach Ag Business landowners with a conservation message at different points in time that may move them into a different landowner category having a stronger connection to the land.

AG BUSINESS	ASSET OWNER	CITATION
Land-use pragmatists: Primarily view forests as economic resources. Recognize a trade-off between economics and conservation and make it clear that their priority is profit and not conservation.	**Indecisive:** Not sure about how to best manage their land but value timber as a source of income and investment. Utilize property for consumptive recreation.	Goyke et al. 2019 (Southeast)

It is important that policy decisions not disenfranchise those landowners who seem to be lacking a land ethic or seem to express a frontier hero attitude. Instead, we must identify ways in which we can act to foster and expedite the development of their land ethic or sense of stewardship (see "Section VI: Fostering the Stewardship Pathway"). Once they begin a land ethic journey, the process could begin to play an important role in their decision-making and therefore lead to wiser management decisions.

It has been proposed that as people begin to understand their responsibility as stewards of public goods (clean air, water, and healthy landscapes), there will be decreased need for compulsion and coercion in natural resource policy. It is obvious that there is an increasing sophistication trend in the landowner community. Many of today's landowners do understand their impact on the land and take seriously their responsibility to steward it. For this reason, the role of the carrot and stick in natural resource conservation will need strong reevaluation.

By acknowledging distinct groups of landowners and their varying goals, ambitions, ideals, and values, agencies and organizations and their personnel will be equipped to more effectively reach the greatest number of landowners and subsequently impact the largest land area through strategic packaging and promotion of conservation programs to landowners.

CHAPTER 8

Common Motivations

ISRAEL PARKER, ANDREA FELDPAUSCH-PARKER, AND NILS PETERSON

The goal of this chapter is to illustrate some of the most common motivating factors leading one toward stewardship. Case studies presented in this chapter represent how a combination of motivations can be utilized to gain traction with land stewards and to catalyze conservation efforts.

Introduction

Private land management is unique due to its focus on individual landowner goals, objectives, and concerns. The hundreds of thousands of private landowners in the United States and across the globe present a varied tapestry distinctly different from governmental, corporate, or organizational ownership. The sheer number of different owners spread among many different ecosystems, cultures, local histories, value systems, and economies ensures diversity in private landowner priorities for land management and behaviors. In reflecting on this breadth of diversity across the country and globally, a few common motivations and influences have emerged to shape individual landowner behaviors across a variety of land uses. These motivations include and can be categorized into (1) financial considerations, (2) individual versus community responsibilities, and (3) top-down regulations versus opportunities for cooperation. We illustrate these issues in two case studies in this chapter.

Financial Considerations

Whether a rancher on remnant tallgrass prairie, a forest owner in the US Northeast, or a member of a groundwater association, economic realities dictate many stewardship decisions. Financial considerations are a key influence on private landowner behavior. Owners of working lands (generally defined as lands that provide food, fiber, fuel, and/or water) are often reluctant to expend significant funds for land improvements or conservation actions with uncertain benefits when many land uses already provide low revenue. The real or perceived financial burden of stewardship activities such as wildlife habitat improvement can negatively impact landowner participation in or persistence within land stewardship programs. Those programs must fit within the financial framework of individual agricultural operations. Land improvement programs that support agricultural operation budgets and are effectively explained and supported by program managers are most likely to be implemented.

Other chapters of this book delve into various landowner programs, so here we touch on only a few. Financial realities provide opportunities for agencies to engage landowners in mutually beneficial land-use arrangements. *Voluntary Incentive Programs* (*VIPs*), for example, can provide financial assistance to private landowners for certain stewardship practices such as wildlife habitat improvement, reduction of soil erosion, and participation in forest management programs. Landowners express broad support and interest in these types of incentive programs. From a landowner perspective, VIPs lower the capital investment threshold, often provide clear rules and goals that can potentially improve land profitability, and dovetail with widely held stewardship characteristics held by many private landowners (Dayer et al. 2017). As an example, New York State cooperates with the Natural Resources Conservation Service (NRCS), US Fish and Wildlife Service (USFWS), and private landowners to support the Working Lands for Wildlife program, which cost-shares the creation and maintenance of wildlife habitat on private lands in New York.

A particularly long-standing VIP is the *Conservation Reserve Program* (*CRP*), which pays landowners rental to take environmentally vulnerable farmland out of production and establish native plant species. The US Department of Agriculture (USDA) supervises enrollment and management of the CRP. Another regional effort includes the *Regional Conservation Partnership Program* based in the US Farm Bill that partners the NRCS with private landowners to address land stewardship concerns. *Conservation Easements* (*CEs*) are another voluntary program but are generally permanent rather than term-limited like VIPs or CRP efforts. Generally, in exchange for permanently surrendering certain rights to property, landowners receive direct payments or tax incentives, thus providing capital injection for working lands and protecting land from fragmentation or loss. CEs are important tools for conservation, as almost 200,000 CEs covering over 30 million acres have been implemented as of 2020 (National Conservation Easement Database 2020). These arrangements often have overlapping purposes such as food and fiber production, recreation, and species and habitat conservation.

Individual Versus Community Responsibilities

A *stewardship ethos* (i.e., the characteristic spirit of an individual or community as manifested in its beliefs and aspirations) is widespread in private landowner surveys and includes a twofold responsibility to (1) sustain the land and (2) maintain community values in the context of cultural, interpersonal, and familial values for the land (Cook and Ma 2014; Quartuch and Beckley 2014). The vagueness of these goals reflects the individual and contextual meaning of stewardship. Although often outweighed by economic or cultural considerations, many private landowners in the United States and abroad express an interest in balancing conservation with land use in a way that promotes sustainability. *Sustainability* is itself a malleable and contextual term but is at the intersection of environmental, social, and economic issues. Perhaps the simplest way of explaining landowner intentions for their land is to say they generally want to choose the best feasible option that allows continued use and enjoyment of their land. For instance, ranchers surveyed in Utah often viewed ranching as a way of life instead of strictly employment, and this contributed to a determination to hold on to their land even when land prices were high (Cook and Ma 2014). These same ranchers saw themselves as stewards of the land with some responsibility to future generations, their neighbors, and society. This sense of responsibility directly impacted their willingness to work with agencies in initiatives that affected larger swaths of society rather than just their personal property (e.g., carbon sequestration or water quality).

The sense of responsibility to community has distinct limits. Community is often a secondary concern behind loyalty to family and future generations (Peterson et al. 2010). Put another way, the maintenance of the land for current and future family outweighs the needs of neighbors and community. Peers can still promote conservation by establishing norms for participating in wildlife management activities (e.g., creating deer habitat) or providing communication networks for sharing conservation knowledge (e.g., landowner groups). Landowners, however, may differ in following their neighbors in some cases. While some research has shown that social influence can encourage landowner engagement in conservation actions, this should not be treated as a universal response. Byerly et al. (2019) found that landowners were less likely to engage in certain actions if informed that neighbors were already undertaking those actions. This could reflect the belief that the neighbors' efforts would be enough to accomplish community goals.

We need to continually revisit foundational concepts such as *stewardship* and *autonomy*. Even though most land in the United States is privately held, *private property is a malleable concept* because laws, demographics, and economics change over time. Thus, the idea of proper care of the land is not just relative to local realities but will change in those local areas. For instance, private land stewardship may change along with demographics as Texas sees shifts in Hispanic ownership with roots in Mexican private property culture (Peterson et al. 2010). Additionally, the relationship between public and private ownership may change. The Alberta Land Stewardship Act (2009) in Alberta, Canada, for example, redefined provincial land-use planning to emphasize incentives and market-based conservation initiatives (MBCIs). Within this new legislation, however, was the ability to incentivize regional plan directives on private lands in a variety of ways (Roth and Howe 2011). Although private land ownership remained, the underlying expectations of that ownership appeared to change in light of a new regional authority.

Regulation and Cooperation

The willingness to engage broader issues on an individual level comes with distinct caveats. Quartuch and Beckley (2014) state that *regulations and incentives often motivate landowners.*

Most landowners, however, prefer incentives over regulations because incentives are less likely to erode their autonomy, complicate decision-making, or impact their livelihood. Although incentives in the form of crop subsidies largely define decision-making for many farmers, subsidies may be viewed as an erosion of autonomy among some landowner groups. The strong ribbon of private property rights ties many landowners together in the United States and produces a pronounced desire to avoid or minimize regulation from local or national governments. Power flows from this collective desire to protect private property rights and adds important political considerations to any discussions regarding the issue. This is not to say that private landowners want the absence of regulation. Their desire for stewardship seems to dovetail with the reality that some regulations provide a net communal good as long as they are reasonable, fair, and consistent.

Cooperation with Landowners

Private lands are far more extensive than public lands in the United States, making engagement with landowners central to effective natural resources management. The current landscape of property ownership in North

America emerged from centuries of negotiations between nation-states and individuals. The negotiations, and contracts emerging from them, range from unjust and immoral practices (e.g., various takings of land from Native peoples) to the social contracts implicit within the US Constitution protecting citizen landowners from property seizure and occupation. The former type of relationships led to the overuse of natural resources during the 18th and 19th centuries across much of the continent (e.g., clear-cutting of forests), resulting in the extirpation of many previously abundant wildlife species. Learning from this history, cooperative relationships may not only build more effective relationships between landowners and governments, but they may also provide more sustainable solutions by leveraging sustainable practices landowners have adopted over time. Because property is operationalized as a relationship between the state and individuals (e.g., states enforce private property rights, and landowners pay taxes and allow legal trespass), it makes sense that natural resources management is most effective when landowners are engaged as collaborators (Quartuch and Beckley 2014). This is particularly important if agencies desire the long-term success that requires landowners to continue management measures even after agency support ends. Further, political power inherent in collective private property rights concerns can provide legislative support for cooperation, such as through wildlife tax valuation practices.

Collaboration between landowners and agencies is complex and needs nurturing to create and maintain. The quest to find common ground among a multitude of landowners and agencies requires some level of *shared values* about the system and an agreement that private land is an important component and contributor to the public and the sustainability of larger ecosystems. For instance, private lands are critical for endangered species conservation; however, landowners commonly fear that finding endangered species on their land will lead to decreased private property rights (Olive and McCune 2017). Furthermore, many landowners may have little interaction with government officials or knowledge about endangered species conservation. This makes it difficult to leverage the general support that private landowners have for nature into broad and enduring management actions. Thus, natural resource managers and agencies must make consistent efforts to engage private landowners across multiple areas of concern. Discussions about endangered species habitat management will likely revolve around private land autonomy, financial hardship, community values, and agency support. This subject will often invoke fears about erosion of property rights and loss of autonomy. To be effective, these discussions should focus on the landowner's goals and the intersection of those goals with preferred endangered species management practices. Ultimately, endangered species conservation success is based on developed relationships among landowners and agencies or organizations that provide lasting financial incentives, personal satisfaction, and other economically driven opportunities.

Third-Party Support and Landowner Groups

Often partnerships between private landowners and government agencies benefit from a third-party "honest broker." Nongovernmental organizations (NGOs) such as the Environmental Defense Fund (EDF) and The Nature Conservancy (TNC) may serve as intermediaries in several key ways. For example, EDF managed endangered species credit markets for golden-cheeked warblers in Texas. Other NGOs such as land trusts promote and manage voluntary conservation easements that have permanently protected wildlife and agricultural uses on millions of acres of private land. In these contexts, landowners can protect critical wildlife habitat when resources at state or federal wildlife management agencies are not available. Other examples include Pheasants Forever, Ducks Unlimited, and the National Wild Turkey Federation that partner with USDA to engage and support landowners in providing information and technical expertise for wildlife habitat conservation and restoration projects.

Focused collaborations can provide mutually beneficial agreements. For instance, in some prime white-tailed deer habitat in Texas, revenue from hunting leases exceeds the agricultural production value of the land. This has led to the creation of multi-landowner groups within specific areas that share common goals of deer habitat management. These landowner associations provide alternative revenue sources for landowners, contribute to habitat maintenance when conducted properly, and prevent urban subdivision. Though these are often landowner led, they include collaborations with state agency support and advice.

The overarching landowner preference is for collaboration in finding mutually beneficial and long-term land management strategies. Benoit et al. (2018) found that ranchers outside of Calgary, Canada, held four basic landscape values as important: (1) lifestyle and community values, (2) ecological values, (3) production values, and (4) economic values and property rights. This largely mirrors private land research in other locations and indicates a need for agencies, municipalities, and conservation organizations to clearly understand these values as they seek collaboration and support with private landowners.

Examples of Private Land Stewardship in Action

The following case studies look at specific instances of private land stewardship that characterize these three motivating factors for private landowners. In our first case study we delve into the multiparty management of an endangered species on private lands. Our second case study illustrates the complex relationship between hunting views and wildlife management.

CASE STUDY 8.1

Private Landowner Participation in Endangered Species Management for the Federally Endangered Ocelot

Ocelots are a small, solitary cat species that range from the southern tip of Texas (occasionally occurring also in Arizona) through Mexico and into Central and South America. Though they can weigh as much as 35 pounds, they are generally twice the size of a house cat, with tawny hues and brown spots with elongated black borders. They subsist on a generalist diet, feeding on small and medium-sized mammals, birds, lizards, and fish. This diverse diet allows them to survive in a multitude of habitats ranging from the flat thornscrub of southern Texas to the tropics and mountainous regions of South America. Ocelots are considered a species of concern by wildlife managers because fewer than 100 exist in the United States. Low numbers and threats to individuals and habitat led to their addition to the Endangered Species List in 1982 via the Endangered Species Act (ESA). According to a 2014 study by Janecka et al., "habitat fragmentation, resulting from human induced changes to the landscape during the 20th century, created island populations with highly reduced gene flow and increased genetic drift and inbreeding."

This reduction in genetic diversity makes the population less resistant to environmental stressors and less reproductively successful and thus has the possibility to lead to extinction in the United States. Ocelots are also protected through the Convention on International Trade in Endangered Species of Wild Fauna and Flora, also known as CITES. Currently they can be found on one national wildlife refuge in Texas (Laguna Atascosa National Wildlife Refuge), as well as neighboring private lands. It should be noted that well over 90% of Texas is privately owned (Texas Land Conservancy reports about 95%), with the rest made up of local, state, and federal lands. This means that to effectively manage ocelots in the United States, the US Fish and Wildlife Service (USFWS) needs to work with private landowners to save habitat and connect populations.

In the United States, protection of endangered species has always been a contentious issue due to disagreements over appropriate wildlife management strategies, property rights, and regulatory power. Organizations such as the Environmental Defense Fund (EDF) have attempted to alleviate some of the tension by working with private landowners in a manner that respects their sense of individualism while concurrently protecting critically endangered Texas species such as the black-capped vireo, golden-cheeked warbler, Houston toad, and ocelot. As an international nongovernmental organization (NGO), EDF has a mission to "preserve the natural systems on which all life depends," using science, economics, partnerships, and nonpartisan policy. In the case of the ocelot, EDF chose incentive-based programs to preserve habitat with the assistance of state and federal agencies and other NGOs as partners. Specifically, the Landowner Conservation Assistance Program (LCAP) and Safe Harbor Program (SHP) were created in 2002 and 2005, respectively, to restore and enhance areas in the Texas Rio Grande Valley. These habitats support the last remaining ocelot population in the United States, and these programs were modeled after other Conservation Assistance and Safe Harbor Programs. For example, the golden-cheeked warbler and the black-capped vireo programs in Central Texas enrolled 33 landowners in the LCAP within the first three years of the program. The LCAP for the ocelot provides landowners with technical assistance in addition to a cost-share for restoration efforts on their property. In 2008, an addition to the LCAP was made, called the State Acres for Wildlife Enhancement (SAFE) practice for thornscrub restoration. SAFE allows landowners to enroll in the Conservation Reserve Program (CRP) using revegetation with thornscrub as an approved CRP practice according to the Farm Bill. The SAFE program requires a 10- to 15-year contract and provides landowners with a sign-up incentive payment in addition to annual rental payments and maintenance costs. The SHP focuses on the restoration of ocelot habitat without the risk of possible regulatory burdens placed on Case landowners through the ESA. This program allows landowners to be placed under EDF permits with the USFWS.

In 2009, an evaluation of EDF's efforts was conducted roughly a year after launching the program. This evaluation sought to identify and prioritize factors that could bolster landowner participation in habitat restoration

FIGURE 8.1A, B. Michael Tewes teaching his graduate students, Jennifer Korn (camo) and Arturo Caso (black) (photo courtesy of Larry Ditto). **FIGURE 8.1C.** Researchers at the East Foundation taking measurements of a tranquilized ocelot (photo courtesy of © Sharon Wilcox).

CASE STUDY 8.1 (*Continued*)

Private Landowner Participation in Endangered Species Management for the Federally Endangered Ocelot

and conservation activities for the endangered ocelot in Texas, a state known for strong private property rights. Conclusions were based on interviews with landowners and EDF program partners (i.e., USFWS, Texas Parks and Wildlife Department, TNC, NRCS, Farm Service Agency [FSA], and the Valley Nature Center).

From the interviews conducted, it was determined that trust and respect for personal value systems played a major role in landowners' willingness to participate, in addition to the monetary benefits (Feldpausch-Parker and Peterson 2009). Serving as an intermediary between the USFWS and the public, EDF was able to incorporate other goals landowners may have had for their properties, such as enhancing land for the purposes of hunting/wildlife viewing or creating a vegetative border between properties. EDF also allowed for more flexibility in their programs, catering to the needs of landowners in addition to wildlife. Finally, EDF was able to effectively disseminate program information via their partners' communication systems, including NRCS and FSA newsletters and citizens active in the local nature center, as well as by word of mouth and news media. The main limiting factor in program recruitment was the fact that EDF was targeting specific tracts of land for revegetation, ruling out participation by willing landowners outside the designated area. The rate of exurban development in the area was also an issue since development was either planned or under construction in areas needed for the corridor.

CASE STUDY 8.2

North Carolina Sentinel Landscapes Partnership

North Carolina is a unique mix of military installations, agriculture, natural resources, and rapid urbanization. Land managers and planners have long understood that the natural friction inherent in conflicting land uses can lead to poor decisions and lose-lose situations. It can also undermine long-term planning as competitors fight for the primary dwindling resource: land. This can be seen with increased urban encroachment around military installations, which reduces mission success, decreases the number of farms and overall farmland, and leads to rapid deforestation. The North Carolina Sentinel Landscapes Partnership (NCSLP; established 2016; North Carolina Sentinel Landscapes 2021) is a product of the realization that cooperation and planning can best address these complex issues and provide multi-stakeholder benefits in the state of North Carolina.

North Carolina State University administers the broad partnership that includes Marine Corps Base Camp Lejeune, US Forest Service, Environmental Defense Fund, North Carolina Soil and Water Conservation Districts, and North Carolina Farm Bureau. This cooperation directly addresses the three landowner motivations discussed earlier: (1) financial considerations, (2) individual versus community responsibilities, and (3) regulation and cooperation. The NCSLP develops "economically driven solutions" that support military installations by enhancing working lands and natural resource management. This has required close cooperation among stakeholders, leading to millions of dollars in funding for initiatives such as (1) Voluntary Agricultural Districts (VADs) that empower local agricultural landowners. VADs are localized efforts to maintain agricultural lands by providing relief from some municipal assessments, due process prior to condemnation, and other benefits.

FIGURE 8.2. The 2d Battalion, 6th Marine Regiment participates in live-fire exercises at Camp Lejeune, North Carolina.

(2) Market-based conservation initiatives leverage easements into land protection and agreements between local agricultural operations and military installations. These efforts may include term-limited restrictions on activities on agricultural lands to minimize impacts on military operations. Fair market prices are determined that are acceptable to both the military and local landowners. (3) Food and Fuel 4 Forces are dedicated to creating a pipeline between local agricultural producers, military installations, and power producers. Local agricultural products could provide fresh food and renewable fuel for local initiatives. These efforts maximize local cooperation and community engagement to make all stakeholders part of the solution to ongoing issues.

The NCSLP is based on the premise that conservation of natural resources is in everyone's best interest. As buffer lands are protected around military installations, these facilities have increased mission flexibility, such as training airspace and mobility exercises. Local, nonmilitary landowners receive financial incentives for participation without regulatory overhead. These financial considerations help protect small farms and ranches, conserve local wildlife resources, and protect natural resources such as timber and water that support rural and urban areas alike. It is important to remember that urban areas are highly dependent on the ecosystem services that rural areas provide, such as clean water, clean air, flood control, recreation, and food. Protection of rural lands safeguards urban viability and benefits rural landowners.

Stakeholder cooperation will increase in importance as urban land uses continue to push into rural areas. The intersection of competing land uses need not necessarily lead to poor outcomes, however. With cooperation and planning, preferred outcomes such as preservation of rural livelihoods, protection of military installations, and provision of ecosystem services are possible. The onus is on partnerships that leverage multiple funding sources directed toward stakeholder-agreed programs.

SECTION IV

The Practice of Stewardship

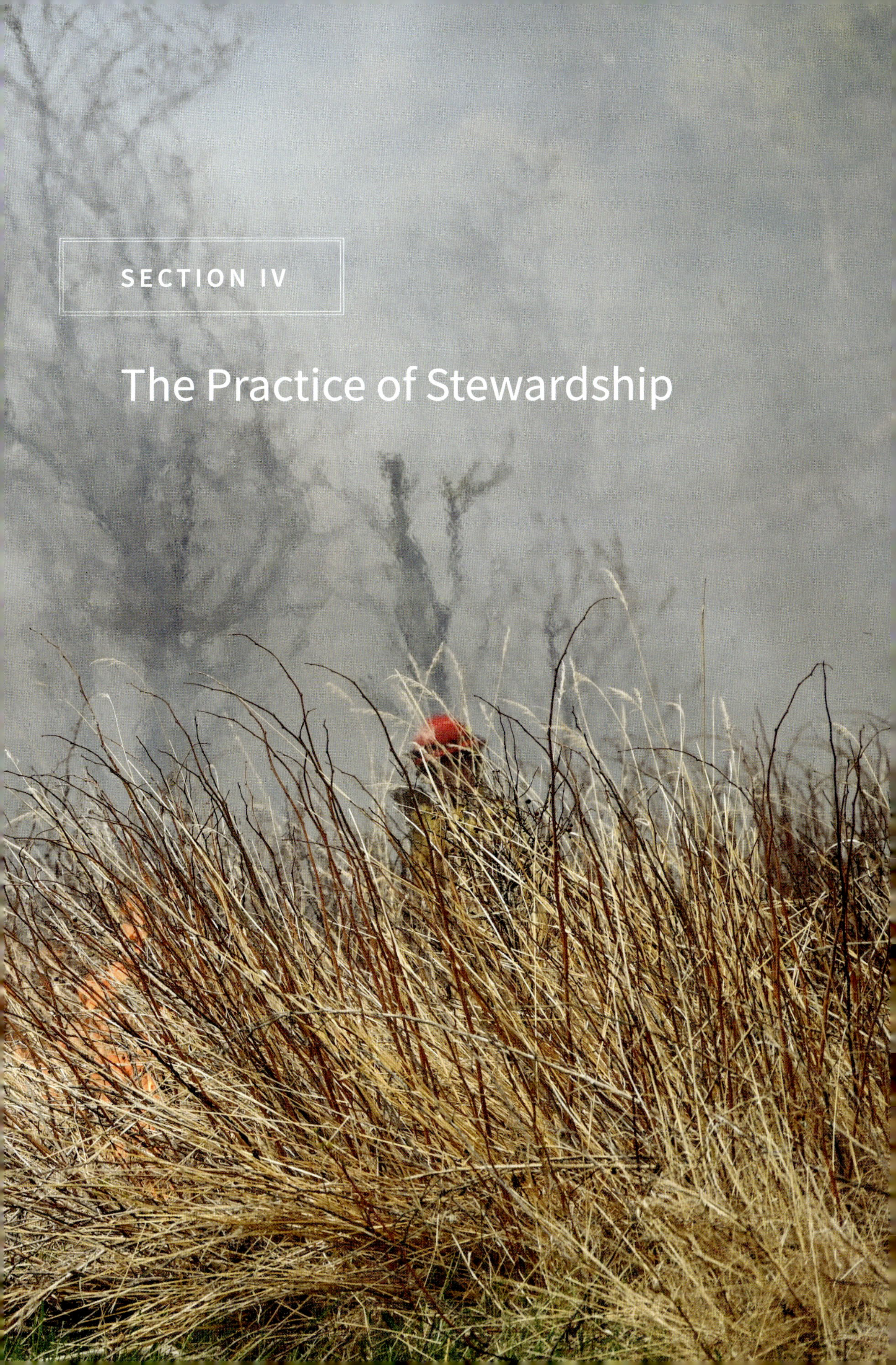

CHAPTER 9

Approaches and Tools

JAMES C. CATHEY, ABIGAIL H. HUDDLESTON, AND BRUCE BEARD

Approaches to land stewardship in the United States have evolved significantly in the last several centuries from those of frequent overconsumption to more harmonious techniques that seek to balance use, appreciation, and preservation of natural resources. This chapter discusses influences and perspectives that can shape a landowner's approach to private land stewardship, the history of land stewardship as the federal government attempted to balance public and private interests, and some current tools and programs available to assist landowners in their private land stewardship endeavors.

Introduction

In a relatively short period of time, from the colonization of the Eastern Seaboard to the exploration and settlement of the West, the predominant views on the relative scarcity of natural resources have evolved significantly. In the late 1800s and early 1900s, mechanization and agriculture played a pivotal role in drastically altering landscapes as American colonists and generations that followed intensively worked the land to the point where some farms were near collapse. Since those early years, it has become more common to incorporate purposeful, *comprehensive* land stewardship practices into farm and ranch plans, but because native wildlife and plant species have been viewed as competition to growing crops or livestock predators, there is ingrained skepticism among many traditional agriculturalists toward this movement. Even so, the tide is shifting. Interest in nature continues to grow, and in many areas of the nation land is being purchased for its recreational or inherent value rather than its agricultural production value. Today, families with multiple generations who have operated on the same parcel of land and individuals who are newer landowners are all discovering a deeper connection to nature.

Land Stewardship Influences: Louis Bromfield

The agrarian lifestyle was certainly altered alongside the landscapes of the 1930s Dust Bowl, as soil was poorly managed and lost through wind and water erosion. Management practices had to change to save the soil and agricultural value of these lands before they were lost forever. This is perhaps what inspired Pulitzer Prize–winning novelist Louis Bromfield to purchase the Ohio-based Malabar Farm in 1939, with the sole purpose of experimenting with new farming practices that would restore the land. Novel techniques such as contour plowing, planting cover crops, reforestation, and fencing cattle out of wooded areas were used to increase the fertility of the land. Ultimately, the property served as a teaching farm, demonstrating to others the benefits of these techniques that are now commonly used throughout the country (Ohio's Country Journal 2017). Farmers have always had a strong tie to the land, but the success of Bromfield's Malabar Farm was a stark illustration of the value of improving land health through true stewardship. Much like Bromfield's position as a leader and visionary in agricultural stewardship, Aldo Leopold was a figurehead for landowners interested in nature who wished to apply agricultural tools in new ways to restore wildlife habitat.

Land Stewardship Influences: Aldo Leopold

Whether a stewardship journey begins with an individual, family, or cooperative, there is always a catalyst that pushes an individual to take the first step toward making management decisions that promote the long-term health of the land, thereby improving wildlife habitat and preserving ecosystem services. Modern-day steward Dick Cates of the Cates Family Farm in Spring Green, Wisconsin, shared, "My dad gave me a copy of [Aldo Leopold's] *Sand County Almanac* for Christmas when I was 16 (figure 9.1), and that started a lifelong conversation about the importance of stewardship and caring for your part of the earth and being a

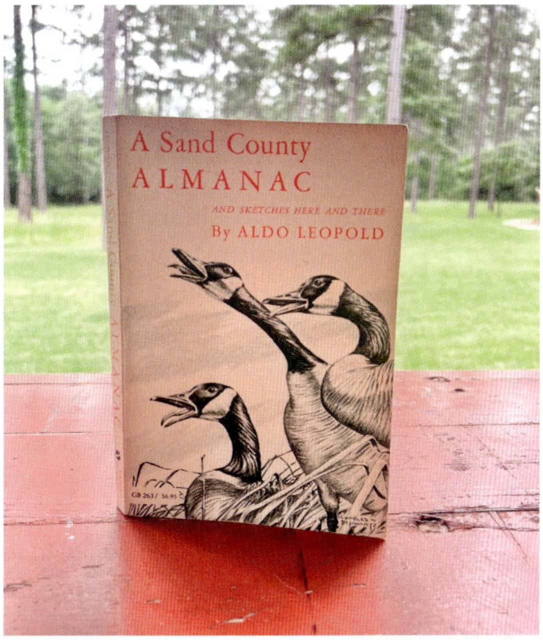

FIGURE 9.1. *A Sand County Almanac.*

productive member of the community." Land stewardship can be complex but persists because individuals desire to leave property in much better condition than they found it. Land stewards like Aldo Leopold, an author, naturalist, philosopher, and conservationist who is considered by many to be the father of wildlife management, have a deep sense of responsibility to improve their property and its natural resources for others to enjoy far into the future.

Consumptive land management practices like hunting for population management, not nonconsumptive practices like prairie restoration, are often the default system in the United States and evolved from European influence. Leopold's book *Game Management* describes this shift while establishing a scientific approach to modern-day wildlife management. This book firmly established the profession of wildlife management and how to apply the scientific method to observations of the outdoors. As the body of knowledge grew, it spawned specialty areas that drive academic programs across the nation today. Leopold noted that sometimes applying small changes in key areas at the right moment yields extraordinary outcomes. Today these actions are most likely to be accomplished by the private landowners who apply the tools of Leopold—axe, cow, plow, fire, and gun—to create healthier habitat and wildlife populations with each new management effort.

Although the growing interest in natural resource management is commendable, there are varying views of private property among landowners (Yung and Belsky 2007). For instance, many states and private landowners advocate for a process where complete control of the land is the responsibility of the landowner (Brick and Cawley 1996). Others may agree with this concept but recognize that local customs require landowner cooperation to accomplish common goals. In western states, cooperation among ranchers is required for maintaining fences and equipment, handling livestock (e.g., branding, capture, care, trailing among properties), and managing emergencies like wildfires or grassland weed encroachment. This collaborative effort among landowner communities protects their livelihood as well as their way of life. Individualistic mind-sets are set aside for the common good of the greater community. Social obligations like these became customs over time, but accelerating changes in land ownership and the views of new landowners often challenge established ideas. New owners may not feel compelled to follow local norms or have the necessary background in livestock management or other necessary techniques, leading to disagreements within communities. Given the independent nature of most landowners, it seems that communication is needed between federal and state governments, ranchers, and wildlife enthusiasts alike to recognize common areas of interest that beget the healthy lands needed for their interests to thrive.

Historical Approaches to Land Stewardship in the United States

The United States has a long, complicated relationship with land and has been both praised for its preservation of national parklands and wildlife habitat and deeply criticized for its confiscation and destruction of Native lands and private property. The US government has had various approaches to private land stewardship since the nation's founding, ranging from indirect funding of conservation to intense regulation. The Founders advocated for the protection of white men's property from tyrannical government rule and often equated the ownership of land with personal freedom. President John Adams declared, "Property must be secured or liberty cannot exist." This argument was enunciated through the passage of the Fifth Amendment to the US Constitution, which states: "No person shall . . . be deprived of life, liberty, or property, without due process of law; nor shall private property be taken for public use, without just compensation." Although there is debate among the federal government, Indigenous peoples, and private landowners on the meaning of "just compensation," as discussed in the following chapter, this is one early example of the government using its legislative powers to reduce the chances of property owners being stripped of their rights.

The Acquisition of Private Lands for Public Purposes

Despite the US government's desire to protect and work with private landowners, one of its main approaches to land stewardship has historically been acquiring land to create protected areas for wildlife habitat, recreation, or national parks. In 1832, the federal government created the Hot Springs Reservation on native Quapaw Indian land in Arkansas. This was the first federal land set aside for recreational use and habitat protection and would later be designated as a national park (NPS 2015). In 1849, the Department of the Interior (DOI) was established to control these newly acquired lands, and within the DOI, the US Fish and Wildlife Service (USFWS) was created in 1940 (DOI 2020a). When President Theodore Roosevelt was inaugurated in 1901, he became a champion of this public land movement and is often referred to as the "conservation president." He used his presidential authority to create the US Forest Service (USFS) in 1905, established 150 national forests and five national parks, and protected approximately 230 million acres of public land during his tenure (NPS 2017).

Technical and Financial Assistance

The US government has a vested interest in public land management but has also designed specific organizations to cooperate with private landowners and provide them with the technical and financial assistance they need to steward their land. The Soil Conservation Service (SCS) was created in 1935 under President Franklin Delano Roosevelt. Later called the Natural Resources Conservation Service (NRCS), this agency was established at a time when natural resources were used unsustainably and landowners were in desperate need of both the knowledge and resources to restore the land. For example, during the Dust Bowl era in the 1930s, Texas farmers struggled with topsoil loss resulting from the common farming practices of the day. With the technical assistance of the Civilian Conservation Corps (CCC) and the SCS approximately 5,000 Texas farmers learned the practice of contour plowing and terracing their fields to prevent precious topsoil from washing away (Ramos 2008).

In addition to direct landowner assistance and education, another federal approach to private land stewardship that has generally been well received is financial assistance for landowners interested in implementing conservation practices. The Land and Water Conservation Fund (LWCF) of 1964 is a massive USFWS effort to partner with states and territories to provide both economic incentives and cooperative endangered species fund grants for landowners practicing sustainable forest management. The Forest Legacy Program has protected approximately 2.37 million acres of forestland and provided $1.4 billion in funding, and the Cooperative Endangered Species Conservation Fund consists of 246 grants totaling $394 million in funding (USDA 2026; USFWS 2026). As of March 2020, the funding for the LWCF was made permanent by the Great American Outdoors Act (S.3422), ensuring a strong future for these conservation efforts and particular funding sources.

Present-Day Approaches to Land Stewardship

While it is difficult to maximize short-term agricultural production revenue and ecological health simultaneously, a stewardship approach that seeks to optimize these values for the long term will benefit both the land and landowner for generations. The conservation experiment of the United

States is ongoing, and across the vast expanse of the nation's lands, there are variations in values, beliefs, and management programs.

Most states have both state and federal entities that help landowners achieve their management goals, and although there is some overlap in expertise, each is specialized enough to accommodate the variable needs of landowners (table 9.1). Often landowners turn to one or more of the following groups: NRCS, Cooperative Extension units, state or federal forest services, and/or their state or federal department of natural resources. NRCS works with landowners in a more holistic approach to conserve soil, water, and natural resources and will provide landowners with technical guidance to develop a ranch management plan. Cooperative Extension approaches agricultural activities from the land-grant mission point of view, where knowledge gained through research is translated into practical application for production agriculture. State and federal forest service groups keep the production of timber and natural forests in a healthy and sustainable condition, while state/federal departments of natural resources are focused on wildlife and their habitats.

TABLE 9.1. Summary of federal and state conservation funding programs by agency

Acronym	Program	Agency*	Description
CELCP	Coastal and Estuarine Land Conservation Program	NOAA	Competitive state and local grants to acquire property or conservation easements from willing sellers within a state's coastal zone or coastal watershed boundary.
CESCF	Cooperative Endangered Species Conservation Fund	USFWS	Grants to private landowners and groups to implement conservation projects for listed species and at-risk species. Funded activities include developing Habitat Conservation Plans, land acquisition, habitat restoration, research, and wildlife management.
CFOSCP	Community Forest and Open Space Conservation Program	USFS	Matching grants (50/50) for local governments, tribes, and nonprofit organizations for full-fee purchase of forestlands; differs from Forest Legacy in its community focus and the requirement of fee purchase plus public access. The program is part of the 2008 Farm Bill; no appropriations yet.
CRP	Conservation Reserve Program	NRCS	Pays farmers to take environmentally sensitive cropland out of production and plant long-term resource-conserving covers (e.g., grasses and trees). The CRP offers 10–15-year contracts with annual rental payments, incentive payments for certain activities, and cost-share assistance to establish approved ground cover on eligible cropland in order to reduce erosion on sensitive lands, improve soil and water, and provide significant wildlife habitat.

TABLE 9.1, CONTINUED Summary of federal and state conservation funding programs by agency

Acronym	Program	Agency*	Description
EQIP	Environmental Quality Incentives Program	NRCS	Technical assistance, cost-share payments, and incentive payments to assist crop and livestock producers with environmental and conservation improvements. The EQIP, established by the 1996 Farm Bill, offers financial, educational, and technical help to install or implement structural, vegetative, and management practices and is one of the several voluntary conservation programs that are part of the USDA "Conservation Toolbox."
FLP	Forest Legacy Program	USFS	Competitive grant program and direct payments to support land acquisition (fee purchase and easement) to protect important scenic, cultural, fish, wildlife, and recreation resources as well as riparian areas.
FPP	Farmland Protection Program	NRCS	State and local grants to help purchase easements that would preclude nonfarm development of productive farmland.
FSP	Forest Stewardship Program	USFS	Provides technical and educational assistance to nonindustrial private forest owners to develop forest management plans.
HFRP	Healthy Forests Reserve Program	NRCS	Restores private forest ecosystems to (1) promote the recovery of threatened species, (2) improve biodiversity, and (3) enhance carbon sequestration.
LIP	Landowner Incentive Program	USFWS	State grants to protect and restore habitats on private lands to benefit at-risk species (including federally listed, proposed, or candidate species).
LWCF	Land and Water Conservation Fund Federal Land Acquisition	NPS	Program to acquire new federal recreation lands in cooperation with the National Park Service.
LWCF	Land and Water Conservation Fund State and Local Assistance Program	NPS	State matching grant program (at least 50%) where states request funds from the National Park Service for specific projects.
MBCF	Migratory Bird Conservation Fund	USFWS	Land and water acquisition or rental as recommended by the secretary of Interior for the protection of migratory bird species. Funding comes from Duck Stamp revenues, import duties on arms and ammunition, and refuge admission fees.

Acronym	Program	Agency*	Description
NAWCA	North American Wetland Conservation Act Grants	USFWS	Matching grants to organizations and individuals to implement wetlands conservation projects in the United States, Canada, and Mexico. Funding comes from congressional appropriations as well as fines and penalties collected under the Migratory Bird Treaty Act of 1918, the Sport Fish Restoration and Boating Trust Fund, and interest accrued on the Wildlife Restoration Trust Fund.
NCWCA	National Coastal Wetlands Conservation Grants	USFWS	Matching grants to states for acquisition, restoration, and enhancement of coastal wetlands. Funding comes from the Sport Fish Restoration and Boating Trust Fund, which is supported by excise taxes on fishing equipment, motorboat and small engine fuels, and import duties.
NERRS	National Estuarine Research Reserves System	NOAA	Grants to coastal states to acquire lands and waters necessary to ensure long-term management of an area as a national estuarine reserve and for operations, construction, and education programs.
REPI	Readiness and Environmental Protection Integration	DoD	Cost-sharing program for the acquisition of easements from willing sellers as a way to preserve high-value habitat and limit incompatible development around military ranges and installations.
RTP	Recreational Trails Program	DOT	Apportionments (maximum 80%) to the states to benefit outdoor recreation, including hiking, biking, inline skating, and equestrian use.
SFR	Sport Fish Restoration Program	USFWS	Apportionments to the states for fishery projects, boating access, and aquatic education. This program is funded by the Sport Fish Restoration and Boating Trust Fund, which is supported by excise taxes on fishing equipment, motorboat and small engine fuels, and import duties. The annual allocation to the Sport Fish Restoration Program from the Dingell-Johnson Act or the Wallop-Breaux Act is equal to 57% of the trust fund's receipts (after annual deductions).
SWGP	State Wildlife Grant Program	USFWS	Grants to plan and implement programs that benefit wildlife and habitats, including species not hunted or fished. Funding comes through appropriations from the Land and Water Conservation Fund.
UCFP	Urban and Community Forestry	USFS	Forest-related technical, financial, research, and educational services to local government, nonprofits, community groups, and educational institutions.

TABLE 9.1, CONTINUED Summary of federal and state conservation funding programs by agency

Acronym	Program	Agency*	Description
WRP	Wildlife Restoration Program	USFWS	Apportionments to the states to restore, conserve, and manage wild birds and mammals and their habitat. This program is funded by the Wildlife Restoration Trust Fund, which is supported by excise taxes, authorized by the Pittman-Robertson Act, on hunting equipment.

Note: DoD, Department of Defense; DOT, Department of Transportation; NOAA, National Oceanic and Atmospheric Administration; NPS, National Park Service; NRCS, Natural Resources Conservation Service; USFS, US Forest Service; USFWS, US Fish and Wildlife Service.

The Endangered Species Act and Related Programs

Presently, there are varying approaches to land stewardship, and some landowners benefit financially from government assistance, while others fund management entirely themselves. Perhaps this choice of how landowners advance land stewardship is dictated by trust (Peterson and Horton 1995). Some landowners want nothing to do with state or federal employees, and others have a history of great relationships and experiences with employees from every level of government. One recent approach to land stewardship, the passage of the Endangered Species Act (ESA) in 1973, was considered a great victory among conservationists but has since caused much more regulatory concern for landowners than other financial incentives (Kohm 1991; Bean 1998). Landowner apprehension of regulation creates a dilemma for the USFWS, as the successful management of threatened and endangered species relies on successful management conducted on private lands (Bean and Wilcove 1997; James 2002). Knapp et al. (2015) suggested that ESA regulation may be efficient on federal lands but may lead to negative consequences on private lands, where private landowners often perceive the fiscal and legal requirements of threatened and endangered species conservation as too burdensome (Sorice et al. 2011; Ciuzio et al. 2013). Policy changes including Habitat Conservation Plans (HCPs), Safe Harbor Agreements (SHAs), and Candidate Conservation Agreements with Assurances (CCAAs) have improved conservation of many species and trust among landowners, but these tools are still viewed with skepticism (Evans et al. 2016; Henson et al. 2018). There is room for continual trust building between governmental organizations and landowners, and both sides have much to gain. For instance, CCAAs allow landowners to participate in voluntary conservation activities that benefit species of concern and ultimately seek to prevent listing. If a species is eventually listed, the USFWS has committed to participating landowners that they will not be subject to additional federal restrictions of the ESA. This agreement gives the landowner greater assurances of their land and water use, and the agreement becomes an automatic permit for some level of incidental take of the species of concern if listing does occur (Epanchin-Niell and Boyd 2020).

NRCS Programs

Some of the most recognized conservation incentive programs are those offered by the NRCS, which often has a favorable relationship with landowners, who value its strong agricultural connection and prevalence on the local level (table 9.2). The most recent Farm Bill, the Agriculture Improvement Act of 2018, encompasses a number of agriculture and food programs like commodity support, conservation, trade and food aid, nutrition assistance, farm credit, rural development, research and Extension activities, forestry, energy, horticulture, crop insurance, livestock, agriculture and food defense, and historically underserved producers. This omnibus multiyear law that extended to 2023, with subsequent one-year extensions in 2024 and 2025, is the primary funding source for NRCS conservation programs that aid landowners in their efforts to improve land health. Dean Jackson, owner of a third-generation dairy and row-crop operation in Pennsylvania, attributes his land ethic to "a very good conservation district in our area. Nathan Dewing, with the Bradford County Conservation District, knows our story inside and out. He has worked very hard to build a relationship with the people of Bradford County, and that's where everything starts—[building relationships] with farmers." This local expertise and investment in personal relationships allow programs such as the Environmental Quality Incentives Program (EQIP) and Conservation Stewardship Program (CSP) to successfully invest private landowners in conservation. EQIP provides agriculture producers with financial assistance and one-on-one technical guidance from experts to implement improvements on their land, and through the 2018 Farm Bill update, now provides eligible historically underserved participants in the program with advance payments so they can offset initial investment costs. Similarly, CSP, which is currently the largest conservation program in the United States, provides both technical and financial assistance to landowners but is targeted to those with preexisting conservation plans that can be improved by identifying the natural resource problems within the operation. These NRCS programs help improve land health, create habitat for wildlife, and support landowners in their agriculture operations. Proper

TABLE 9.2. USDA Natural Resources Conservation Service and Farm Service Agency conservation programs (2018)

Acronym	Program	Description
ACEP	Agricultural Conservation Easement Program	Helps landowners, land trusts, and other entities protect, restore, and enhance wetlands, grasslands, and working farms and ranches through conservation easements.
AMA	Agricultural Management Assistance Program	Helps agricultural producers manage financial risk through diversification, marketing, or natural resource conservation practices. NRCS administers the conservation provisions while USDA's Agricultural Marketing Service and Risk Management Agency implement the production diversification and marketing provisions.
CIG	Conservation Innovation Grants	Awards competitive grants that drive innovation and develop the tools, technologies, and strategies for next-generation conservation efforts on working lands. Grantees leverage the federal investment through matching requirements. Through CIG's new On-Farm Trials, partners provide incentive payments to producers to offset the risk of implementing innovative approaches.
CREP	Conservation Reserve Enhancement Program	Encourages landowners to sell or lease long term to beginning, socially disadvantaged, and veteran farmers and ranchers willing to implement sustainable practices or transition to organic production by providing two years of additional payments for expiring CRP-enrolled land.
CRP	Conservation Reserve Program	Protects soil, water quality, and habitat by removing highly erodible or environmentally sensitive land from agricultural production through long-term rental agreements.
CSP	Conservation Stewardship Program	Helps agricultural producers maintain and improve their existing conservation systems and adopt additional conservation activities to address priority natural resource concerns. Participants earn CSP payments for conservation performance—the higher the performance, the higher the payment.
EQIP	Environmental Quality Incentives Program	Provides financial and technical assistance to agricultural producers to address natural resource concerns and deliver environmental benefits, such as improved water and air quality, conserved ground and surface water, reduced soil erosion and sedimentation, and improved or created wildlife habitat.
HFRP	Healthy Forests Reserve Program	Helps landowners restore, enhance, and protect forestland resources on private lands through easements and financial assistance. Through HFRP, landowners promote the recovery of endangered or threatened species, improve plant and animal biodiversity, and enhance carbon sequestration.
RCPP	Regional Conservation Partnership Program	Promotes coordination between NRCS and its partners to deliver conservation assistance to producers and landowners. Under partnership agreements, NRCS and its partners leverage and target their respective resources to deliver conservation assistance to producers and landowners to address priority natural resource concerns.
VPAHIP	Voluntary Public Access and Habitat Incentive Program	Provides state and tribal governments with funding or incentives to expand or improve habitat in existing public access programs.

Note: Most programs require the development of a conservation plan. For further specifics regarding plan requirements and eligible practices, visit USDA Natural Resources Conservation Service, http://www.nrcs.usda.gov.

stewardship of private lands benefits all Americans, as it restores clean air and water, reduces soil erosion, improves soil health, increases plant diversity, and provides wildlife habitat while simultaneously supporting the sustainability of the agricultural enterprises that feed and clothe our nation.

State Programs: Tax Incentives and Project Loans

In addition to these federal programs, conservation incentive programs have become a priority for many states. These programs can vary across locations and offer different assistance levels. In Texas, landowners may receive a property tax rate lower than that of market value for land that is actively being managed to support wildlife habitat. This strategy is known to promote stewardship practices that benefit all Texans through clean air and water, healthier wildlife habitats, and recreational opportunities available on private land. In Wyoming, the state legislature established the Wyoming Wildlife and Natural Resource Trust (WWNRT) program in 2005 with the primary purpose of enhancing the state's natural resources and conserving wildlife habitats. Some of the most common project types funded include prescribed fire treatments to preserve aspen habitats, enhancement of wetlands and other water sources, and invasive species eradication. This large-scale state program has provided funding for projects on over 2 million acres of land, collaborated with more than 130 organizations, and implemented almost 800 projects across the state (WWNRT 2020).

Joint Ventures

State and federal governments, individuals, and conservation groups all have different approaches to conservation projects, and it is gratifying to see examples of cooperation between all three. Since the late 1980s, the USFWS has advanced migratory bird joint ventures where these collaborations work to improve wildlife habitat at the landscape scale. Currently, there are 18 joint ventures that cross state and international boundaries in North America. Through group planning, conservation projects are prioritized to gather data on various conservation needs, address those needs through habitat management, monitor those actions over time, and share the results with others through education and outreach efforts. Partnerships are key to the success of joint ventures and extend the USFWS conservation budget, where the return on the federal investment is $31 for every dollar invested, including nonfederal partnership contributions. Even though all partners are important, including the private landowner is paramount to the overall effort, and with this key constituent and nearly 6,000 other partners, 27 million acres of migratory bird and associated wildlife habitat have been restored.

State Universities

State universities are another trusted source that landowners consult for reliable, scientific information. The Cates Family Farm in Spring Green, Wisconsin, is a perfect example of this partnership and how it can successfully transform land. This vertically integrated grass-fed beef operation, run by Dick Cates, was initially purchased by his father in 1960. A trout stream runs through the center of their pastureland with a main branch about 10.5 miles in length. They grazed cattle on approximately 20% of this stretch of stream, which became a point of contention particularly in the 1980s, when ranchers were pressured to eliminate all grazing in riparian areas. Cates partnered with the University of Wisconsin–Madison to conduct research on this trout stream and determine the effects of grazing on its health and productivity. Through this partnership, the Cates family was able to conclusively determine that their grazing strategy was actually having a positive effect on the stream. Dick Cates shared, "When I was growing up, Lowry Creek was a Class 3 stream, which means it was classified as 'impaired water' by the Wisconsin Department of Natural Resources. And now, this stream that we have grazed livestock along for all the years I've been here (52 years) and managed it for the last 32, will be classified this fall as a Class 1 trout stream—the highest classification for a cold-water stream in the Driftless Area of Wisconsin! The stream is now one of three champions for wild trout propagation for the state." This stream transformation is one of many success stories that originate from partnerships between landowners and public universities, and this success is now shared with the surrounding Wisconsin community through stream-bank workshops and a farmer-led watershed protection group. This single trout stream is promoting the conservation of working lands and uplifting the idea that production agriculture and conservation can coexist.

Conservation Partnerships and Nongovernmental Organizations

Navigating between both ecological and political communities is sometimes difficult but not impossible. America's Longleaf Restoration Initiative (ALRI), an effort that began in 2010 involving landowners, private industry, nongovernmental organizations (NGOs), universities, and state and federal agencies is a great example of collaborative conservation programs with positive outcomes. The once extensive longleaf pine (*Pinus palustris*) ecosystem covered 93 million acres, ranging from Virginia in the north, to Florida in the south, and Texas in the west. Aggressive harvest, fire suppression, and replacement with alternative pine species reduced historic longleaf forests to only 3.2 million acres (ALRI 2021). With the dramatic decline of this ecosystem, we may never really know the loss of biodiversity, but it currently supports about 900 plants, 36 mammals, 170 reptiles and amphibians, and 100 bird species. Among these groups, 30 species are federally listed as threatened or endangered, and others are being considered for listing under the ESA. The goal of ALRI is to increase the number of longleaf pines from 3.2 million to 8 million in 15 years by utilizing rangewide conservation plans and voluntary land management, thereby protecting valuable habitat for vulnerable species as well as native biodiversity. Through innovative partnerships, cost-share funding to encourage voluntary restoration, and management of longleaf by private landowners, the initiative is gaining momentum. By encouraging and supporting prescribed burning, a necessary part of longleaf pine management, the initiative has enhanced over 12 million acres of existing longleaf and restored 1.3 million acres of longleaf stands to date.

Just as ALRI is a collaboration of many partners, other organizations like Ducks Unlimited (DU) work with multiple groups to assure healthy waterfowl populations and their habitats (DU 2020). Organizing and prioritizing the biological needs of multiple waterfowl species are complicated processes, and DU must consider things on an extensive, continental scale as they work to make positive conservation improvements for migratory birds. With projects in all 50 states, Canada, and Mexico, DU's operation relies on a network of relationships involving local landowners and chapters, business owners, conservation leaders, policymakers, and more. DU has influenced policy language in the North American Wetland Conservation Act (NAWCA) and the 2018 Farm Bill to help fund waterfowl habitat projects. NAWCA has reportedly contributed over $1.8 billion toward the conservation of 30 million acres since its inception 30 years ago. These larger efforts act as a catalyst for the actions of local landowners who participate in cost-share programs that promote waterfowl habitat, clean air and water, nutrient cycling, and carbon sequestration that benefits the health of all. DU promotes waterfowl conservation, encourages restoration of working lands, and looks for common ground where many diverse groups can benefit from the improvement of a single parcel of land. For some landowners, working with an NGO like DU reduces the apprehension of interacting with state and federal agencies and offers a local pathway to achieving conservation goals that benefit waterfowl and many other animals and plants. Wetland habitats are ecologically valuable and are understandably a focal point for management, but these bottomlands are also connected to uplands with forests and grasslands whose health improves the entire system. Even though ducks may be used as the rallying cry, it is clear that entire ecosystems are benefiting from the inspiring partnership between DU and private land stewards.

Private Landowner Communication Networks

Not all conservation efforts take place on such a large scale as Ducks Unlimited, but smaller operations have advocates that are equally passionate about their cause. A common trend throughout these groups is a renewed interest in communication networks that reunite neighbors, encourage taking management risks, and invite new, diverse groups to join the conversation. For example, the Blackfoot Challenge stemmed from a recognition by Montana residents that their precious resource, the Blackfoot River, and its associated habitat and fish communities were being compromised by misguided mining, logging, and grazing practices on both private and public lands and jeopardized by residential and commercial development in the area (Blackfoot Challenge 2021). Smaller conversations among a few concerned individuals eventually led to the official organization of the Blackfoot Challenge, where private landowners, business owners, and public agencies now collaborate "to conserve and enhance natural resources and the rural way of life in the Blackfoot watershed for present and future generations." Their charge encompasses 132 miles of river, the 1.5 million acres that form its watershed, and the population of about 9,000 residents within the watershed. This is a lot to take on for any one person, group, or agency, but by having a community-based, inclusive approach to conservation, their improvements to the Blackfoot River basin have also improved their rural way of life (figure 9.2).

Much like the grassroots development of the Blackfoot Challenge, the Malpai Borderlands Group (MBG), located in the desert habitat joining Arizona and New Mexico, created a communication network for the common good. Ranching and grazing lands in the area were being threatened by ongoing fragmentation of larger ranches and subsequent land development. Private landowners held much of this landscape, but federal control of some public lands within the nearly 1-million-acre area led to a common feeling of distrust for the government among local people. However, identifying common goals through a couple of key events and conversations led to the ongoing conservation of this large landscape. Brush encroachment into desert grasslands limited grass and livestock production and compromised the habitat of many wildlife species. Over time, government agencies that were once charged with extinguishing the wildfires that prevented brush encroachment realized its many benefits to both ranchers and wildlife. This

FIGURE 9.2. Blackfoot Challenge group in Helmville, Montana, learning about soil health.

mind-set change created an avenue for common ground among groups. Involvement by state and federal agencies, universities, and especially private landowners has prevented habitat fragmentation using research, private land management, and the implementation of conservation easements held by The Nature Conservancy. The sum of these efforts was a significant success for the MBG, and their management practices bolster about 30 threatened and endangered species found in the region, while also improving range conditions for livestock.

Conclusion

It is humbling to think back on some of the early pioneers of the conservation movement and the contributions they made that continue to uphold the efforts of present-day land stewards. How can a single individual create such substantial change as did Teddy Roosevelt, Louis Bromfield, and Aldo Leopold? Maybe the secret lies within a place like Malabar Farm, Leopold's Shack in Wisconsin, the majestic Blackfoot River, or the harsh beauty of the Malpai Borderlands, where the soul connects to nature on a deeper level and understands that without nature humankind cannot exist. The growing number of both large- and small-scale conservation endeavors is exhilarating, ranging from independent landowners utilizing state and federal incentive programs or just applying their own intellect and perseverance, to land stewardship champions who connect neighbors and launch conservation actions that expand far beyond their own landscape. Jay Tanner, a fifth-generation cattle rancher who is a leader in greater sage-grouse (*Centrocercus urophasianus*) restoration in Utah and manages a small stream that is home to the threatened Lahontan cutthroat trout (*Oncorhynchus clarkii henshawi*), is passionate about bringing other ranchers along with him in his conservation efforts. Tanner said, "I don't think I'm that much different than other good ranchers. Maybe I've been willing to stick my neck out a little further by trying new and different things, but I always try to get my neighbors involved as well. For the most part, that has been a really positive experience." Land stewardship leaders like Tanner are an important catalyst for community-level change and frequently lead to the formation of conservation communication networks. Perhaps true wealth is attained when investments are made in the land—the infrastructure that sustains all life. A movement is taking place among landowners who, in the context of Leopold's land ethic, recognize they are not the conquerors of the land, but instead that the land provides life, and as a part of this ecological community, their participation in its conservation is critical. Although this realization is heartening, the landscapes and plant and animal communities they safeguard are in desperate need of improvements. Although past and present legislation has bolstered conservation efforts, the imagination and ingenuity of the private landowner should, and often do, reach beyond the realm of government-sanctioned conservation. The success of most species cannot be manufactured solely within public lands. Private lands are key to the successful management of wildlife and their ecosystems, so now is the time to advance policies and communication organizations that pursue the improvement of these invaluable public goods on private lands.

CHAPTER 10

The Nexus of Public Lands in a Private Landscape

BRUCE BEARD, R. TOM DARDEN, BUDDY HUFFAKER, MARK MADISON, ABIGAIL H. HUDDLESTON, AND ANDREA E. MONTALVO

With case studies provided by Todd Nightingale, Jamie Sowell, and Lucas Cooksey.

In a book about private land stewardship, why would we have a chapter on public lands? The reality is that in many—primarily western—states private land stewards work closely with public land agencies through grazing leases, cooperative agreements, and habitat and species recovery projects. For this reason, it is important that practitioners are familiar with the mission, focus, and history of these federal agencies and how they may affect private land stewardship.

Introduction

Stewardship of land in the United States is largely dependent on private landowners, but the interactions between private and public lands and their management creates an important nexus that many times yields support structures for landowners, drives policy, and shapes land stewardship outcomes. In the United States, the federal government manages roughly 640 million acres or about 28% of the 2.27 billion acres of land in the country (Vincent et al. 2014). Federal agencies under the umbrella of the US Department of the Interior (DOI) and the US Department of Agriculture (USDA) Forest Service collectively manage approximately 673 million acres of land that are often leased to the private sector for grazing and oil and gas development. Additionally, the US Department of Defense (DoD) manages 28 million acres of military bases, training ranges, and other military-focused lands.

Publicly owned lands are governed by many statutory guidelines that are often vague and occasionally contradictory. Today there are over 100 treaties, international agreements, federal statutes, executive orders, and federal regulations that direct wildlife management on federal lands.

Understanding the institutional context of each agency that controls the stewardship of public lands is a helpful tool for private landowners who interact with the federal agencies providing them with funding and technical assistance or are on neighboring properties. There has always been intense debate around how public lands should be managed, but their initial purposes were often utilitarian: ensuring future timber and water supplies, providing affordable land for livestock grazing, and providing safety buffers around military bases. There are some exceptions (like the national parks established to ensure the opportunity for wildlife viewing and national wildlife refuges specifically designated to support wildlife populations and their niche habitats), but the modern context of environmental protection and the sustainability movement have been superimposed onto a system of land protection and wildlife management.

Regardless of each agency's unique institutional history, all federal agencies generally have a different set of rules and regulations to follow than private landowners. They must comply with legislative mandates such as the National Environmental Policy Act (NEPA) and the Endangered Species Act (ESA), and when combined with society's increased interest in the environment and the public scrutiny environmental actions bring, public land managers face many stricter demands than individual private landowners.

Conservationist Aldo Leopold acknowledged the need to use natural resources for human survival but also argued that this survival depended on maintaining healthy ecological systems. Leopold began his career as a public land manager with the US Forest Service (USFS) and later became a professor at the University of Wisconsin. After working in both the public and private sectors, Leopold came to believe that while government land management is important for creating sanctuaries, the growth of private land ethics is beyond the power of government. It is on an individual level that society must take responsibility and obligation of caring for the land—a collective community of soils, waters, plants, and animals (Leopold 1949).

This chapter surveys the interactions between federal land management agencies and privately owned lands exemplified in Leopold's own career. Federal agencies hold management power over great swaths of land in the United States, and private landowners who are developing and implementing the principles of a land ethic must be able to successfully work in conjunction with the government. This chapter includes the (1) USFS, (2) US Army, Navy, and Air Force, (3) US Army Corps of Engineers (USACE), (4) Bureau of Land Management (BLM), (5) US Fish and Wildlife Service (USFWS), and (6) National Park Service (NPS). We discuss in depth the mission and goals of each agency and the common ways in which they interact with private landowners.

Department of Agriculture

United States Forest Service

The mission of the Forest Service is to sustain the health, diversity, and productivity of the nation's forests and grasslands to meet the needs of present and future generations. The Forest Service manages 154 national forests and 20 grasslands in 44 states and Puerto Rico that total more than 193 million acres of public multiple-use land. The agency provides technical and financial assistance to state, private, and tribal forestry agencies and is home to the largest forestry research organization, all to achieve multiple uses and benefits and the sustained yield of renewable resources such as water, forage, wildlife, wood, and recreation.

Additionally, the Forest Service provides community assistance and cooperates with state and local governments, forest industries, and private landowners to help protect and manage nonfederal forest and associated range and watershed lands.

Established in 1905, the Forest Service had to face the widespread exploitation and overconsumption occurring in our forests at the time and develop an agency that could sustainably conserve natural resources for future generations. This work is achieved through a network of research units and community and landowner support. Policies like the NEPA of 1970 and the National Forest Management Act (NFMA) of 1976 created pathways for further regulation and preservation of public forests and established the importance of maintaining biodiversity in these ecosystems. A 2012 update to federal regulation to implement the NFMA mandated that the Forest Service is required to do what it can to keep the forest hospitable for all native plants and animals—effectively modernizing the policy to better address 21st-century issues like sustainable recreation, conservation biology, and climate change.

The public is most likely to interact with the Forest Service when visiting a national forest or grassland for recreational purposes, but for the nearly 11 million private landowners who own more than half of US forestland, the Forest Service can be a valuable partner in forest conservation and sustainable timber management. Through a combination of direct technical assistance, economic incentives, and policy implementation, working cooperatively with the states and nongovernment organizations, the Forest Service partners with private landowners to steward our nation's forests.

Programs like the Forest Stewardship Program (FSP), which "provides assistance to owners of forest land where good stewardship . . . will enhance and sustain multiple forest resources and contribute to healthy and resilient landscapes," and the Forest Legacy Program (FLP) that encourages the protection of privately owned forestlands through the creation of conservation easements and strategic land purchases are excellent examples of the work the Forest Service does with private landowners.

CASE STUDY 10.1

Texas

The USDA Forest Service strives to be a good neighbor to the private landowners that surround the four national forests in East Texas and the two national grasslands in the northeastern part of the state. Authority from the Wyden Amendment allows the Forest Service, under certain circumstances, to conduct prescribed burning on private lands to allow for increased safety and efficiency and help meet management goals for federal forestlands and adjacent private lands. For example, if a private property is surrounded by a Forest Service property (an "inholding") or the private ownership is between a roadway and the Forest Service boundary, the Forest Service may partner with the private landowner through an agreement to conduct the prescribed burning on the private property in conjunction with the planned Forest Service burn. This would be of no cost to the landowner, and the burn would be conducted by Forest Service personnel as part of the larger burn. When allowable, utilizing this authority can be mutually beneficial to each party by, for example, (1) assisting private landowners with getting their property treated with prescribed fire by professional prescribed burning staff; (2) making for more efficient use of federal resources to accomplish their burning missions to manage national forests and reduce hazardous fuels; (3) creating safe alternatives on perimeter firebreaks when roads, waterways, natural fuel breaks, or more efficient routes can be utilized; (4) benefiting neighboring properties by reducing hazardous fuel loadings that help mitigate devastating wildfires; and (5) improving wildlife habitat across boundaries.

This across-boundary cooperation promotes prescribed fire on the landscape and is mutually beneficial to both private landowners and public interests. The collaboration between neighboring properties to find solutions to mutual challenges has many benefits beyond these examples. These activities foster positive relationships and create networks to increase the pace and scale of prescribed fire in fire-adapted systems and promote a positive culture around prescribed fire. These opportunities for cooperation, information sharing, and demonstrating best practices promote landscape-level forest improvements by meeting multiple management objectives and creating resilient and adaptable landscapes across boundaries for the benefit of future generations.

Department of Defense

Army, Navy, Air Force

The mission of the DoD is to provide the military forces needed to deter war and to protect the security of our country. The DoD manages 28 million acres of public land in the United States for various military uses across more than 4,800 defense sites ranging from parcels of less than 1 acre to the 3.1-million-acre Nellis Air Force Range in Nevada (Vincent et al. 2014). Underlying every management decision affecting DoD lands is the requirement that these lands must be managed first to support military training and testing—a situation quite different from that of other agencies that were created solely for the purpose of traditional land management and conservation.

> All DoD natural resource conservation program activities shall work to guarantee DoD continued access to its land, air, and water resources for realistic military training and testing and to sustain the long-term ecological integrity of the resource base and the ecosystem services it provides. DoD shall manage its natural resources to facilitate testing and training, mission readiness, and range sustainability in a long-term, comprehensive, coordinated, and cost-effective manner. (DoD Instruction 4715.03 of March 18, 2011, cited in Vincent et al. 2014)

On many military installations, limiting public access due to security considerations and the need for safety buffer zones has protected their natural resources from development and degradation, resulting in many healthy, intact landscapes. One key tool the DoD uses to address threats of incompatible development and habitat loss around military installations is its "buffer authority" (10 U.S.C. § 2684a). This provision was enacted in 2002 and authorizes the DoD to partner with conservation organizations and state and local governments to promote land uses that are compatible with the military mission and preserve high-value habitat around installations. The promotion of farming, ranching, and forestry and the preservation of open spaces around military installations and beneath military airspace benefit wildlife and rural economies, provide recreational opportunities, and forestall damaging developments such as wind turbines and high-density housing.

In 2013, the DoD, USDA, and DOI, in conjunction with the White House Rural Council, established the Sentinel Landscapes Partnership to identify opportunities and eliminate obstacles to coordinating their programming on private lands of mutual interest. Through this partnership, the departments

CASE STUDY 10.2

Georgia

In 2011, the US Fish and Wildlife Service was evaluating the need to add the gopher tortoise to the federal threatened and endangered species list. Federal listing could further restrict military operations at numerous key military installations across the species range from North Carolina to Florida. In a proactive response, a new DoD initiative, which included the DoD, Military Services, US Fish and Wildlife Service, representatives from each state Fish and Wildlife Agency, and other species experts, was formed to help conserve gopher tortoises alongside preserving the military training options in the Southeast. The outcome of this initiative was the Department of Defense Conservation and Crediting Strategy whereby the military can earn credits for land acquisition, habitat restoration, and other approved efforts to conserve gopher tortoises and their habitats on protected areas. Those credits could then be used to offset impacts from military training if the tortoise was federally listed in an area where the impacts occur. Another goal of this strategy was to help aid in the acquisition and conservation of lands supporting gopher tortoise populations sufficient to help preclude the need for federal listing of the species at all. One application of this strategy occurred when the Georgia Department of Natural Resources acquired the Alapaha River Wildlife Management Area (WMA) in 2016 with support from the US Army, the USFWS, and other nonfederal partners. The 6,900-acre WMA east of Tifton has nearly 2,000 gopher tortoises—the highest number and greatest density on any state-owned tract. Alapaha River WMA also has plenty of other at-risk species, such as gopher frogs, striped newts, eastern indigo snakes, and pond spice.

This case study is relevant to private landowners because it utilized the "buffer authority" of the DoD thru 10 U.S.C. 2684a to provide an avenue by which the Georgia Department of Natural Resources worked with a willing private landowner to sell the property for this purpose. Additionally, it required communication with other private landowners around the change to inform them it was now a publicly managed recreational area. This in turn spurred some additional opportunities for the local public in the form of a new destination recreational area where goods, services, and equipment rental opportunities were created.

While the catalyst of this was the DoD and support of the military mission, it took numerous public agencies, private landowners, and now continued interaction and support from the general public to make the Alapaha River WMA a success. This success also included the gopher tortoise as intended, where the USFWS recently determined in 2022 that the species was being sufficiently conserved throughout the range to no longer require consideration for federal listing.

recognize and reward landowners within these "sentinel landscapes" with incentives for land practices that insulate military lands and meet the objectives of any applicable conservation programs. The partnership connects landowners with assistance opportunities that meet their land management needs while insulating military lands, such as the Agricultural Conservation Easement Program (ACEP), Agricultural Management Assistance (AMA), and Candidate Conservation Agreements (CCAs).

The Sikes Act of 1960 is the primary authority under which the DoD manages natural resources on its installations. In 1997, the Sikes Act Improvement Act updated the original law to require most military installations with significant natural resources to develop Integrated Natural Resource Management Plans (INRMPs) in cooperation with the USFWS. A further amendment in 2008 authorized military installations to enter into cooperative agreements with other parties to maintain and improve natural resources located off military lands. This is good news for private landowners who live on properties neighboring military installations—through the Sentinel Landscapes Partnership and the Sikes Act, they can partner with the DoD to incentivize managing private lands to benefit the military mission as well as conserve natural resources and wildlife habitat.

Army Corps of Engineers

The USACE stewards 12 million acres of public lands and waters through hundreds of water resources projects nationwide. The Water Resources Development Act (WRDA) outlines the environmental operating principles of the Corps and acts as the primary governance over Corps activities. The WRDA is typically renewed by Congress every two years and deals with various aspects of water resources, including environmental, structural, navigational, flood protection, and hydrology. This legislation, in conjunction with the Flood Control Act of 1970 and Estuary Restoration Act of 2000, grants authority to the USACE to address environmental issues and conduct environmental restoration projects relating to water resource planning. Although the USACE is not subject to the Sikes Act, its Lakeshore Management Plans function in much the same way as INRMPs. As a federal agency, USACE also has an affirmative requirement to manage its land and actions in accordance with applicable federal, state, and local laws. These Lakeshore Management Plans help articulate methodologies for doing just that.

One interaction the private landowner may have with the USACE is through floodplain easements—if a private property falls within a historical floodplain, it may be subject to flood mitigation management by the agency. Additionally, if a landowner's property contains a flowage easement or neighbors public lands under a flowage easement, the USACE owns the right to flood the land in perpetuity. Additionally, through USACE's navigational management requirements both public and private landowners are engaged from time to time in support of dredging and channel maintenance operations.

CASE STUDY 10.3

Lower Mississippi River

In 2013, U.S. Army Corps of Engineers (USACE) Mississippi Valley Division (MVD) released the *Conservation Plan for the Interior Least Tern, Pallid Sturgeon, and Fat Pocketbook Mussel in the Lower Mississippi River* (Channel Improvement Program [CIP] Conservation Plan), which outlines a process to conserve the three endangered species within the footprint of the CIP in the Lower Mississippi River (LMR). The Service conducted a consultation under Section 7(a)(2) of the ESA with the USACE on their conservation plan.

The consultation process required close inter-agency communication and collaboration during all phases of development. The USACE's ongoing implementation [includes] basic research and data gathering on the species and their habitats, channel engineering design and testing, modeling, development of Best Management Practices protective of the species and their habitats, and monitoring and analyzing results.

Over the course of the consultation, it became apparent the very programs (e.g., the CIP) that most significantly affected the endangered species and their river habitats could be important and effective tools to maintain and enhance its ecological functions. This resulted in the identification of conservation opportunities that could be effectively incorporated into existing channel improvement or maintenance projects, with little to no additional program costs, and with no negative impact to the USACE's primary flood management and navigation safety missions. It also resulted in significantly improved habitat and population baselines for all three endangered species within the LMR portion of their ranges. The CIP consultation demonstrates that numerous benefits for the species, the ecosystem, and the agencies can be derived under Section 7(a)(1). . . .

The USACE Conservation Plan, strategies and actions to maintain and enhance ecosystem complexity of the LMR include significant benefits for recreational uses of the river. For example, secondary channels, dike notches, and dike tips provide spawning and feeding areas for multiple game and forage fish species, and provide the most productive fisheries habitats in the Mississippi River. Secondary channels also provide safe fishing, paddling, and wildlife viewing opportunities outside of the navigation channel. (Excerpted from *USFWS Fact Sheet*, September 2014)

It is important to remember that large-scale projects such as this occur both on and adjacent to private property owners, and, reading between the lines, this effort required immense coordination with the landowners adjacent to these navigational channels, people who utilize the river every day for both personal and commercial activities, and with locations needed to place the dredged spoils from the CIP Program. It is relatively easy to see the navigational dredging and construction operations taking place but likewise easy to forget that these vast projects touch all sizes, shapes, and types of private lands along the LMR unless you yourself are one of those landowners. Again, the overall success of the project was the collaboration of federal, state, nongovernmental, and private landowners to make implementation of the CIP and Conservation Plan happen in a way to sustain the river system for flood control and commercial activities alongside meaningful conservation for species and habitats.

Department of the Interior

Bureau of Land Management

The BLM in the US DOI is responsible for managing 261 million surface acres of public land and 700 million acres of subsurface minerals. The BLM manages more plant and wildlife habitat than any other federal or state agency in the nation. The Federal Land Policy and Management Act of 1976 (FLPMA) gave the agency a defined mission, provides for administration of public lands on multiple-use basis, and requires land-use planning that includes public involvement and a continuing inventory of resources. Managing public lands for multiple-use management aims to utilize the various resource values of public lands in the combination that will best meet the present and future needs of the American people. When authorizing land-use activities such as recreation, livestock grazing, energy development, and forest management, the BLM must ensure that the needs of wildlife, fish, and plants are taken into consideration. Today, the agency also works to improve the health of entire watersheds to sustain and enhance a variety of biological communities.

One of the most prolific public and private landowner interactions is the use of grazing leases. On 155 million acres of the land managed by the DOI, the BLM administers nearly 18,000 grazing permits and leases to ranchers to graze their livestock on over 21,000 allotments. In 2017, the BLM announced a new initiative known as Outcome-Based Grazing Authorizations (OBGAs) that established more collaboration between the BLM and livestock ranchers when making grazing decisions. This initiative improves grazers' flexibility to respond to changing conditions like drought or wildfire and grants the BLM authority to ensure that rangelands and wildlife habitats remain healthy.

CASE STUDY 10.4

Utah

Jay Tanner and his family own and operate Della Ranches in the Grouse Creek Valley of Utah. Primarily a cow-calf operation, the ranch covers an area of about 800,000 acres, including 20,000 deeded, 20,000 leased, 55 sections of conservation grazing easements, and the remainder on BLM and Utah State School trust leases. In all the ranch is 90 miles long, 40 miles wide, and ranges in elevation from 4,200 to 9,000 feet.

"In our area, the public, particularly the BLM, owns about 50% of the property. And so, to some degree, I have to work to make sure that the public land managers are doing a good job, and as well, they with I," said Jay of the symbiotic relationship he has with public lands personnel. Since the Sage Grouse Initiative started in 2010, Jay has worked closely with BLM and US Fish and Wildlife Service to promote habitat for the declining species.

"In our area, the public lands have a lot of wildlife habitat, but it's the poorer habitat, it's the drier habitat, in many cases. Most of the riparian areas, most of the water, most of the green groceries for sage-grouse and other wildlife species are on private land. So, a lot of those species of concern might have 50% to 80% of their habitat on that public land, but 50% to 80% of the priority habitat is on private land," Jay noted.

Jay has invested countless hours in recent years removing juniper encroachment across his property, primarily to benefit his cattle, but he eventually realized that it was benefiting the sage-grouse as well. "The wonderful thing was that the sage-grouse started using the areas where we removed the juniper, and it's made the areas significantly better for other species of wildlife as well. And now the BLM folks are doing the same practices where they can on their lands."

In describing how important private use of public land in Utah is, Jay said, "The point is, as a cattle person, a rancher, an agriculturalist, if I can't use the public lands, I go out of business or I have to use my private lands harder. I would end up using them for grazing at a time when wildlife may need to use them as well."

In closing, Jay said the following about the importance of the public-private nexus in a state like Utah: "I think we helped [BLM] to see that on these areas where there's sage-grouse, which has a bunch of private and public lands, that one couldn't function without the other."

US Fish and Wildlife Service

The mission of the USFWS is to work with others to conserve, protect, and enhance fish, wildlife, plants, and their habitats for the continuing benefit of the American people through federal programs relating to migratory birds, endangered species, interjurisdictional fish and marine mammals, and inland sport fisheries. The primary objective of the agency is to guide the conservation, development, stewardship, and management of the nation's fish and wildlife resources. Together with the Department of Commerce's National Marine Fisheries Service, the USFWS is the principal administrator of the federal ESA, which protects federally listed threatened and endangered species of fish, wildlife, and plants and their designated critical habitats. It administers the Migratory Bird Treaty Act, which protects avian species identified in various international treaties. It also manages more than 570 refuges within the National Wildlife Refuge System and 38 wetland management districts totaling 95 million acres. Under its Fish and Aquatic Conservation program, the agency operates 71 National Fish Hatcheries, 51 Fish and Wildlife Conservation Offices, 7 Fish Technology Centers, and 6 Fish Health Centers.

The nature of the USFWS mission requires the agency to work with others to advance the conservation of fish, wildlife, plants, and their habitats. Their Partners for Fish and Wildlife Program "provides free technical and financial assistance to landowners, managers, tribes, corporations, schools, and nonprofits interested in improving wildlife habitat on their land." Any landowner can request free consultations on how to improve their land stewardship practices and receive the free technical and financial assistance needed to implement their management plans. The program will also connect private landowners with partnering state and federal agencies and nongovernmental organizations to manage their private properties.

National Park Service

The mission of the NPS is to preserve the natural and cultural resources and values of the US National Park System for the enjoyment, education, and inspiration of this and future generations. The National Park System includes over 400 areas covering more than 84 million acres in every state, the District of Columbia, American Samoa, Guam, Puerto Rico, and the Virgin Islands. These areas have been deemed of such national, historical, scenic, and scientific significance as to justify special recognition and protection in accordance with various acts of Congress. At the heart of the NPS's mission is the mandate to conserve natural spaces for present and future generations of people and wildlife.

One obvious intersection between private landowners and the NPS is the use of the National Park System for its recreation and entertainment value, but according to the NPS, more than 2.6 million acres of privately owned lands are located within the park boundaries placed by the National Park System. Depending on the use of the lands, the Land and Water Conservation Fund provides federal funding to acquire some of these privately owned properties to protect their resources and provide for visitor use and enjoyment. The NPS may reach out to landowners to purchase either their lands or interests in them or conduct a land exchange.

CASE STUDY 10.5

Alaska

Privately owned lands within the boundaries of national parks are known as inholdings and are the result of lands that predate the designation of a park. Generally, owners of inholdings are allowed to use their properties in a manner similar to that of other private property owners in their state. However, they may be subject to additional federal and agency-specific regulations regarding access to their homes and use of their lands.

Within national park units in Alaska, there are more than 1.6 million acres of land owned by private individuals, corporations, the State of Alaska, and local governments. These owners and other valid occupants are entitled to adequate and feasible access to their property. However, the process of obtaining legal access across NPS lands has never been well described in a single document.

Upon receiving numerous comments from individuals, corporations, and governments, in addition to holding public meetings, the NPS issued a policy document for landowners to better understand the process of authorized access across park areas. The User's Guide to Accessing Inholdings in a National Park Service Area in Alaska describes how landowners and other valid occupants can secure rights under Section 1110(b) of the 1980 Alaska National Interest Lands Conservation Act (ANILCA). It addresses motorized access to inholdings, as well as access that would require improvements such as construction or maintenance of a road, power line, landing strip, or similar undertakings on publicly owned NPS lands.

It is not always feasible and/or necessary to acquire privately owned lands within a park's boundary. However, many of these lands are important for the visitor experience and resource preservation interests. Therefore, private landowners play an essential role in the protection of natural and cultural resources, and their stewardship helps preserve our nation's heritage found in our national parks for current and future generations.

Conclusion

Public lands and their resources have played an important role in American history, contributing to the strength of the federal government; driving economic development; generating the resources like timber, forage, fiber, energy, and minerals on which American livelihoods depend; and providing a platform and framework for natural resource and wildlife conservation. As demands on federal lands have escalated, so have the conflicts between private landowners and the federal government over land uses and values. Some lands, such as those managed by the USFWS and the DoD, have singular missions that make their purpose much clearer to private landowners. Others with multiple-use mandates, such as the BLM and USFS, often endure more conflict with private landowners surrounding their diverse and ever-changing land uses, values, and management strategies. At some level, every federal land management agency is capable of cooperating with private landowners to improve the land health for all, but the resources and aid they provide are often obfuscated by bureaucracy. It is important for private landowners to understand the subtle distinctions between each federal agency, their missions, and their management strategies to find a public/private land interaction niche that benefits them. Understanding the institutional backgrounds of each federal land management agency ensures easier cooperation and understanding between private landowners and federal agency employees who share the mission of land stewardship for wildlife, their habitats, and future generations.

CHAPTER 11

Legal and Regulatory Challenges

TEMPLE STOELLINGER, JIM BRADBURY, AND KARRIGAN BORK

Few natural resource professionals have a strong understanding of the often-challenging legal landscape that can affect the practice of stewardship. This chapter presents some of the most common legal issues to facilitate basic literacy and an appreciation for the potential ramifications of legal actions.

Introduction

Although most public attention is focused on the conservation opportunities and legal challenges on public land owned by the federal government, private land accounts for more than 60% of all land in the United States and thus represents a significant conservation opportunity. However, conservation requires the imposition of limits on human activity, and those limitations are often opposed, particularly when they interfere with the economic productivity of private land or with an established way of life.

In this chapter, we cover a framework of environmental laws that affect conservation on private land, including eminent domain, energy development, the Endangered Species Act (ESA), water rights, conservation easements, and habitat leases. We describe the legal challenges these laws can present and highlight some of the opportunities they offer for landowners engaged in private land conservation. This chapter is designed to introduce private land managers and those who work with them to some of the legal issues commonly impacting private lands.

Fundamentals of Private Property Rights and Conservation

A few fundamentals about private property rights will provide context for the issues in this chapter. Private rights to land, like all property rights, derive from laws enacted by government. State law creates most private property rights, with federal and local law playing complementary roles. The US Constitution safeguards property rights by prohibiting the taking of private property without just compensation. A private property owner may enjoy additional rights as are recognized by statutes or by the common law (law derived from judicial decisions). As these laws change over time, so do private property rights. In short, the government has broad powers to define and regulate private rights in terms of what a landowner can and cannot do.

The government takes many different approaches to achieve conservation goals on private land. Most fall into two categories: prescriptive laws or laws incentivizing conservation. Prescriptive laws define what landowners can and cannot do with their private property. The ESA, for example, prohibits the killing of endangered animals everywhere, including on private land. Conversely, the government provides incentives, usually direct monetary payments or tax benefits, that reward landowners for positive conservation actions. For instance, conservation easements provide an example of an incentive-based conservation approach. They are voluntary legal agreements under which landowners agree to permanently limit uses of their land to protect its conservation values in exchange for a monetary benefit. Beyond these two approaches, others include the creation of markets, government purchase of private land, and educational or technical assistance programs.

When private landowners engage in conservation, they are serving the public interest by conserving public goods such as wildlife and their habitats or ecosystem services like clean air, water, or nutrient cycling. Public goods are defined as a commodity or service whose benefits are shared by everyone but owned by no one. Some of the fundamental questions about private property rights are associated with the conservation of public goods. For example, what rights should landowners have to use their land in ways that disrupt ecosystems and wildlife? What powers should landowners possess, and what limits might society reasonably impose on them?

Water Rights Law and Private Land Stewardship

Water rights law determines who gets to use water and how much they can use. Like most property rights, water rights law is determined at the state level, and most states treat rights to use surface water differently than rights to use groundwater.

Most states in the eastern half of the United States use a type of surface water right called a riparian right. This gives landowners whose land touches a water source, like a river or lake, the right to use water from that source on their land without getting a permit. Historically, the right was limited to what was reasonable, which varied by year and by what other riparian landowners were doing with their water rights. Now, many states use a modified form of riparian rights, regulated riparianism, which means that riparian landowners must get a permit that more clearly defines their water right.

In the western half of the United States, less water is available, and many landowners far from a water source needed a surface water right that would allow them to access water for use on their land. In the mid-19th century, California developed a new form of water right, called an appropriative water right, to address this need. While riparian rights are based on land ownership, appropriative water rights are based on the use of water. Whoever diverts the water first and starts using the water for a beneficial purpose has the right to that water. Water rights that were used first take priority over newer water rights—in a purely appropriative rights system, first in time is first in right. Many states that use an appropriative rights system have modified their system to give more priority to particular water uses, like domestic or municipal use.

Many states use a mix of the two systems for surface water rights. California, for example, still recognizes both kinds of rights. In other states like Oregon and Kansas, the state government ended both systems and embraced a new approach where all water rights require permitting and periodic review of the permits.

States also use a variety of approaches to managing groundwater. In some states, like Texas, landowners can use all the water they are able to pump from wells on their property, even if that reduces the water availability for their neighbors. They can sell that water for use off-property as long as the water is put to a beneficial use. In other states like California, a court will consider the impact of one landowner's pumping on their neighbors when determining how much groundwater the landowner can use. California and Michigan, for example, prioritize on-site use of groundwater over off-site use.

These different systems interact with private land stewardship in a wide variety of ways. Virtually all systems allow water use only for beneficial purposes, and states may or may not consider aspects of land stewardship a beneficial use. Most riparian states require control of water to maintain a water right, so landowners wishing to dedicate their rights to instream flow during all or part of the year may need to seek special mechanisms to do so while also protecting their water rights. Groundwater use to create water holes in low-water areas may be allowed in some states but closely regulated in others.

Overall, states vary widely in the ways they regulate water rights, and these variations may affect the way landowners can use new or existing rights to improve stewardship of their land. Due to the intricacies of water law, decisions about how to manage the water rights associated with private property should be made in consultation with a water rights expert for the state in question.

Eminent Domain

Eminent domain is the power to take private property for a public purpose. Proceedings of eminent domain encroach on arguably one of the most important rights of US citizens: the right to own private property. Property can be taken for a wide array of purposes (e.g., utilities, highways, and railroads) by virtually any governmental entity (federal, state, city, and county) and, in some situations, by private entities (figure 11.1). However, through the Fifth and Fourteenth Amendments of the US Constitution, landowners have the right to receive just compensation for the taking of their property. Most state constitutions likewise afford similar protection against governmental takings without just compensation.

When governments or, sometimes, private entities seek private lands for public use, they generally start by trying to negotiate a purchase from the landowners. If the landowner and the would-be purchaser come to an

FIGURE 11.1. A pipeline scar across a private property in Central Texas (photo courtesy of Lori Olson, Texas Land Trust Council).

agreement, the sale continues like any other sale of private land. However, if the parties cannot come to an agreement, the would-be purchaser can proceed through the eminent domain process by filing a condemnation action in court.

Landowner Compensation Under Eminent Domain

When an entity takes private property through eminent domain, landowners are entitled to just compensation. The payment may include the value of the parcel taken (area value) and the value of the damage to the rest of the landowner's property (remainder damages). However, landowners do not generally recover any attorney fees associated with the process.

The value of the property is determined by one of three approaches depending on the state where the property is located: the Broad Instruction Approach, the Specific Rates Approach, or the Factor-Based Approach. The Broad Instruction Approach gives little guidance on adequate compensation for condemned land besides directions to examine the fair market value of the property. Fair market value is the price recognized between a willing buyer and a willing seller. The Specific Rates Approach, which is used in only a few states, uses statutes to assess a state's interest in property, valuing certain types of land such as heritage, homestead, and agricultural land. These rates are related to the fair market value of the property, but some types of land receive a premium rate. The Factor-Based Approach is a combination between the previous two approaches. States that utilize this approach implement statutes that provide some express terms of property valuation. In addition, if only a portion of the property is condemned, the state will look to see if there was any damage to the property and how characteristics of the property affect the overall value. Some of the factors that may be evaluated are improvements on the property, growing crops, and the goodwill of business.

Landowner Best Practices

The impact of an eminent domain process on landowners involves a multitude of issues beyond compensation. While the process can seem daunting or hopeless to landowners, some of these impacts can be mitigated.

For example, before a petition for condemnation is filed, a condemnor has the right to enter and survey the property. This is an interference with property owners' exclusive use and private enjoyment of their land. Although the condemnor may enter and survey the property, there are limitations to what can be done. Landowners may require a written survey agreement that specifies what may be done on the property, notice before access, hours for access, and restoration of any property damaged by surveying or testing activities.

Landowners facing eminent domain should absolutely consult with an attorney early and often throughout the process. A condemnation proceeding can be costly and time consuming. Although negotiations between the condemnor and the landowner generally must be in good faith, aggressive landmen or condemnors can stoke fear to make landowners feel pressured to accept the first offer to avoid court proceedings. Consultation with an attorney can help landowners not only understand the process of condemnation but also know and understand their rights throughout the process.

Reaching a negotiated agreement with condemnors offers landowners some advantages. While landowners cannot prevent the taking of their property under eminent domain, negotiations often bring positive terms and conditions in the final form of an easement that protects landowners, their property, and their business. An easement is the conveyance document that includes all of the terms, conditions, requirements, and agreements between a landowner and the easement holder (i.e., the condemnor). Landowners can negotiate specific terms that are intrinsically valuable, including limiting the size of the easement, limiting the scope of the easement to one pipeline or transmission line, limiting the ability to enlarge the pipeline or power line, restoring the surface, putting limitations on access to the property, providing for damages for violation of easement terms, and having insurance requirements. Regardless of what condemnors may promise or tell landowners, if it is not in the final written easement, it will not apply. All terms must be included in the final written easement. Landowners must be savvy, not fearful. Seeking out wise counsel and understanding the process and a landowner's rights through the process minimize negative impacts from eminent domain and ensure the continued protection of private land ownership.

Energy Development on Private Lands

Many states use the unified fee ownership theory to define the rights associated with land ownership. Here, the landowner owns everything from the center of the earth to the sky (insofar as they do not interfere with air travel). When a landowner sells or gives away one right associated with a piece of property, we say that those rights have been "severed."

Oil and Gas

In the United States it is common for the mineral rights or interests to be severed from the surface rights on the land. Severance of the mineral interest can occur through a sale (the owner of the land sells just the mineral interest to another party) or through a reservation (the original landowner, often the federal government, or any subsequent landowner sells the property but does not sell the mineral interest). After the severance of the mineral interest, what remains is called the *surface interest*. It is not uncommon for surface interest owners to be unaware that the mineral interest underneath their property is actually owned by someone else. A thorough investigation of a property's title by an attorney or title professional should reveal the divisions of the interests within a property and is always a wise idea when considering a land purchase.

Use of the minerals under a property can affect the owner of the surface interest. For example, ownership of the mineral rights includes an implied easement to use the surface to the extent reasonably necessary to obtain the minerals under the property. The mineral interest owner's easement to use the surface is limited by a standard of reasonableness and an obligation to accommodate the uses of the surface owner if possible. Furthermore, the owner of the mineral interest (who could also be the surface interest owner if the mineral estate has not been severed) has a right to lease or sell the mineral interest. When leasing a mineral interest, the interest owner grants an oil and gas company a leasehold interest, which is a right to develop the oil and gas contained within the mineral interest. Thus, a third party could be the party developing a mineral interest.

Unfortunately, disputes between surface interest owners and mineral interest owners or lessees are a common occurrence, particularly when oil and gas development occurs in populated/urban areas. To address this conflict, many states have enacted surface-owner protection laws. Common elements of these laws include early notice to the surface owner of the proposed drilling operations, the requirement to negotiate surface use agreements, and increased surface damage bonding. While these laws provide surface owners with some additional protections, the mineral estate remains the dominant estate. Thus, the mineral estate owner continues to have an implied easement to use the surface in a manner that is reasonably necessary to develop the minerals. There are ways to practice conservation of surface (i.e., wildlife habitat) and subsurface (i.e., groundwater) resources when a property is developed for oil and gas. New directional drilling technologies allow oil and gas drilling operations to be placed outside of key wildlife habitat or other conservation areas. Surface damage remediation can include wildlife habitat or other conservation efforts. Landowners acting as either a surface and mineral interest owner or just a surface owner can ensure conservation either through their lease with oil and gas companies or through their surface use agreement.

Wind Energy

Some states have found that the right to develop wind can be severed from the surface estate, meaning it can be separately sold from the surface estate, while other states have enacted wind-severance bans. Wind-severance bans ensure that surface owners remain in control of renting and leasing the right to develop wind on their property and to receive royalties or other benefits from its development. However, many of the major wind energy development states—Texas, California, Iowa, Oklahoma, and Illinois—have not banned the severance of wind rights.

One of the major downsides to wind energy development is its impact on wildlife—particularly birds—resulting from habitat loss and fragmentation; electrocution; collision with transmission lines, turbine blades, and towers; and pulmonary barotrauma (which occurs when air-pressure impacts from the spinning turbines cause a bird's lungs to burst). Proper planning and siting of wind farms can reduce the impact to wildlife resulting from wind energy development on private land. Avoiding bird migration routes and important wildlife habitats significantly reduces the negative impacts of wind power on affected species. Wind development on agricultural lands where wind production complements existing land uses provides landowners with opportunity to develop needed renewable energy in a resource-conscious way.

Solar Energy

Solar installations on private land can include large, utility-scale concentrated solar power (CSP) projects or smaller photovoltaic systems mounted on the ground or a roof. Ground-based leases for solar development on private land generally state that developers will not share the leased property with other occupants or even the landowners themselves; unlike wind energy projects that allow for compatible simultaneous uses of the land, solar projects are land-intensive and often require the developer to have exclusive use of the property. Because many solar energy developments are land-intensive, their impact on surface resources can be significant. Some sheep grazing may be possible if the solar panels are sufficiently elevated from the ground. Instead of locating new solar installations in natural environments, many have suggested that solar installations be located on already disturbed or developed land to reduce the environmental impact yet still meet rising demand for renewable clean energy.

The Endangered Species Act and Private Land Stewardship

The ESA has quite a reputation in the private land stewardship community. On the one hand, the act stirs up deep-seated concerns about private property rights, federal regulation of private activities, and ultimately economic consequences. On the other, the act has fundamentally changed the conversation about endangered species and created a strong and enduring public commitment to species protection. Public polling consistently suggests that over 70% of voters support species protection. But there is an interesting twist in the perceived struggle between private property rights and species conservation: Most species protected by the ESA require ongoing active management, and much of that management would, ideally, take place on private land. Because the ESA itself does not require private landowners to manage their land for protected species, and because the agencies responsible for implementing the act do not have the resources to do the kinds of management required on many public lands, those agencies have found ways to interpret the act to encourage private land stewardship. These incentive programs attempt to conserve species while providing relief from the ESA's regulatory burdens and potential penalties. However, these programs can be complicated and understanding them requires some background knowledge on the ESA itself. Despite these complexities, when implemented, incentive programs can contribute to the protection and recovery of species of concern by encouraging landowners to adopt conservation on their properties.

The Structure of the Act

The act protects species listed as threatened or endangered under ESA Section 4. The National Marine Fisheries Service (NMFS) oversees implementation of the act for marine and anadromous species like salmon, while the US Fish and Wildlife Service (USFWS) oversees freshwater and terrestrial species. Species are generally considered for listing after a member of the public submits a listing petition to the relevant agency. Based on the best available science, the agency lists the species as endangered if the species is "in danger of extinction throughout all or a significant portion of its range," or as threatened if it "is likely to become an endangered species within the foreseeable future." Species listed as endangered get the full protection of the ESA, while threatened species get only those protections NMFS or USFWS "deems necessary and advisable to provide for the conservation of such species."

Protections for listed species come from ESA Sections 7 and 9. Section 9 is the heart of the act—it prohibits the "take" of "any endangered species of fish or wildlife." *Take* is a technical term, which the act defines as "to harass, harm, pursue, hunt, shoot, wound, kill, trap, capture, or collect, or to attempt to engage in any such conduct." Both NMFS and USFWS expand this definition to "include significant habitat modification or degradation where it actually kills or injures wildlife by significantly impairing essential behavioral patterns, including breeding, feeding or sheltering." Section 9 applies to everyone—the federal government, state and local governments, private landowners, and members of the public alike—however, it does have some limitations. It does not protect plants on private land, and harming plants on federal land is illegal under the ESA only if it is considered malicious. Section 9 does not require landowners to act affirmatively to preserve listed animal species, and Section 10 provides numerous permits allowing actions that would otherwise violate Section 9.

Section 7 provides most of the act's other protections. It focuses on federal actions, but because many actions by states and private individuals require a federal action, Section 7 can affect those actions as well. For example, wetland draining may require a Clean Water Act 404 permit, and the US Army Corps of Engineers would have to comply with Section 7 before issuing that permit.

Section 7 requires federal agencies to ensure "that any action authorized, funded, or carried out by such agency is not likely to jeopardize the continued existence of any endangered species or threatened species or result in the destruction or adverse modification of habitat of such species." The

CASE STUDY 11.1

State ESAs

Many states administer their own Endangered Species Acts to augment the federal act. These state acts generally mirror the structure of the federal act, although they may add additional requirements. They also tend to offer programs similar to the Safe Harbor Program or other programs to encourage private land stewardship for listed species.

agency undertaking the project (called the action agency) must consult with the relevant expert agency (USFWS or NMFS) to ensure that its action will not violate that requirement. For more significant projects, the expert agency reviews the projects and adds requirements to the project, called reasonable and prudent alternatives, to ensure the proposed project will not jeopardize any listed species or destroy or adversely modify their habitat. Once the expert agency is confident that the project will not violate the law, it issues a *biological opinion*, or BiOp. This describes the project and any required monitoring, allows the action agency to proceed under Section 7, and formally permits (through an incidental take statement) some actions that would otherwise violate Section 9 of the ESA. Under current law, threatened and endangered plants do not require incidental take statements.

In the wetland example used previously, the Army Corps of Engineers would consult with USFWS, and USFWS might require some kind of mitigation for any wetland destruction or seasonal limits on wetland draining. USFWS would ultimately issue a biological opinion setting out these reasonable and prudent alternatives and permitting some Section 9 take of any listed species in the wetland.

Beyond the incidental take statements in Section 7, Section 10 provides additional permits to allow individuals to take species without facing the significant penalties provided by the act. Briefly, Section 10(a)(1)(A) allows take of a species "for scientific purposes or to enhance the propagation or survival of the affected species." Section 10(a)(1)(B) allows take by states and private parties under an incidental take permit, issued in conjunction with a habitat conservation plan that commits the permittee to other actions conserving the species. Finally, Section 10(j) allows reintroduction of populations of listed species as "experimental populations," away from wild populations of the same species, with reduced protections that amount to something like a permit for the public for that particular population. Much of the current ESA administration, from listing and delisting to the outcome of the Section 7 consultation process, is driven by litigation.

Encouraging Private Land Stewardship

Private efforts to conserve listed species are essential. Three quarters of all protected species need private land for habitat, food, or breeding grounds, and more than 90% of listed species occur on private lands. As noted, 84% of ESA-listed species qualify as conservation-reliant, meaning they are not self-sufficient but require ongoing human help. This means that listed species need private land, and they will ultimately need management of private land if they are to avoid extinction.

USFWS and NMFS have developed several programs to encourage private conservation of listed species. The Safe Harbor Program was developed in the mid-1990s as a way to encourage private land stewardship for listed species. In short, landowners agree to improve listed species habitat or forgo activities that would prevent the development of habitat for listed species. In exchange, USFWS or NMWS agrees not to place additional restrictions on the land and issues a permit that allows the landowner to return the land to its baseline condition at some point in the future. This provides at least a short-term benefit to the species and protects landowners' interests in controlling their land. Both the landowner and the relevant agency must approve the final agreement, and it is subject to public comment. As of 2018, 100 Safe Harbor Programs have been approved and cover over 100 million acres and many landowners. Interested landowners can develop these agreements, or join an existing agreement, by contacting their regional USFWS or NMFS office.

Many landowners find opportunities to generate income by providing habitat for listed species. Landowners can establish conservation banks through an agreement with the USFWS or NMFS, which allows them to grant conservation easements to third parties for payments. The easements allow the third parties to compensate for the impacts of their development projects or other projects that destroy listed species habitat, as required under the ESA permits. In the wetland example, a farmer could purchase a conservation easement on a property with a conservation bank agreement, which would mitigate the impacts from draining the farmer's wetland. The farmer gets mitigation credit, the landowner gets payment, and the species gets protected habitat on the landowner's property. Similar agreements are available to mitigate the impact of federal agency actions; these are termed *recovery credit systems*, but their structure mirrors the conservation bank approach. The appropriate regional USFWS or NMFS office can provide more details on both programs. Finally, the Farm Bill provides some tax deductions for private actions that benefit listed species. The USFWS Farm Conservation Program can provide more details on these deductions.

Conservation Easements

In some scenarios, landowners may want to protect their land from future development. One strong tool is the conservation easement, which has become increasingly popular. Entities seeking to comply with the ESA may purchase conservation easements to mitigate any negative impact their actions could have on a listed species or its habitat.

Definition of a Conservation Easement

A conservation easement is a voluntary legal agreement between a landowner and a qualified holder that allows a landowner to permanently limit the type or amount of development on the landowner's property in order to protect the property's conservation and agricultural value while retaining private ownership of the land. Each conservation easement is individually tailored and consists of several elements. These agreements are part of the title of the land, making them binding on all future owners. While the conserved land may be sold or gifted, the land will remain subject to the terms of the conservation easement.

Purpose of Conservation Easements

Conservation easements may be created for a variety of purposes. The Internal Revenue Code recognizes four specific purposes that may be addressed by a conservation easement: (1) preserving land for public outdoor recreation or education, (2) protecting natural wildlife habitats or ecosystems, (3) preserving open-space land, including agricultural land and forests, and (4) preserving land of historical significance or certified historic structures (figure 11.2).

FIGURE 11.2. Conservation easement in Texas protecting open-space working lands (photo courtesy of Texas Land Conservancy).

Conservation Easement Process

There are several types of conservation easements: donated easements, purchased easements, and bargain sale, which is a combination of a donated and purchased easement often used to offset tax considerations. The process is essentially the same for obtaining donated and purchased easements. First, a landowner must determine what rights will be reserved (how many partitions will be allowed, how many buildings allowed, etc.). Second, a landowner must obtain an appraisal, preferably from a professional with experience appraising conservation easements. Third, the landowner will develop a "baseline report" that documents the condition of the property at the time of the transaction.

Conservation easements must be granted to either governmental entities empowered to hold real property interests or to charitable organizations created for conservation purposes. Often conservation easements occur between landowners and land trusts—charitable organizations whose primary focus is land and water conservation. State and federal governments have created programs to hold conservation easements or assist these land trusts in obtaining them. Easement holders own the rights to the easement, granting them the power to perpetually monitor and enforce the easement. The holding entity will typically visit the property once a year to ensure the terms of the easement are being upheld.

Limitations and Restrictions Regarding Minerals and Groundwater

A conservation easement creates restrictions on land use that may greatly limit groundwater and mineral use. The terms of conservation easement agreements typically prohibit the property's groundwater from being transported off-site but do not restrict the use of groundwater on the property itself. Surface mining is prohibited as required by the Internal Revenue Code.

Wetland and Stream Banking Transactions

Regulated by the Army Corps of Engineers and Environmental Protection Agency under the Clean Water Act, wetland and stream banking transactions are a mitigation system of credits and debits that compensate for the loss of wetlands and streams by allocating credits to those who mitigate wetland destruction and giving debits to those who contribute to wetland loss.

Mitigation credits help offset wetland impacts and ensure ecological loss is compensated. Credits are available to public and private entities or to individuals for restored, enhanced, and created wetlands, or they can be transferred to another party to meet mitigation requirements. Credits can also be withdrawn by the owners. This effectively creates a mitigation marketplace to allow efficient wetland protection. Advantages include more numerous perpetual conservation easements and superior ecosystem services at a reduced cost.

Wetland and stream banking transactions are not without their challenges. These transactions require a correct valuation of ecological loss, which can be difficult to assess. Further, these transactions often prohibit future development, effectively functioning as a deed restriction on the property.

Tax Incentives and Implications

Beyond payments for mitigation, other financial motivations encourage conservation easements, such as tax incentives from donated easements. The value of the easement interests donated generates a tax deduction to be used against income. Once the amount of the deduction is known, landowners may annually deduct up to 50% of their adjusted gross income and up to 100% if the grantor is a qualified farmer or rancher. Further, these deductions have a significant carryforward provision allowing landowners to carry forward the deductions for 15 years. There may be additional benefits in the form of property tax savings for landowners who place conservation easements on their land.

Conclusion

This chapter has described several issues of concern for those practicing private land conservation and the complex web of interrelated law, including federal laws not covered here like the National Environmental Policy Act, the Clean Water Act, and the Clean Air Act. More specific state laws add complexity. These laws offer challenges to landowners, limiting what they can do with their lands, but they also offer benefits through the direct monetary incentives or tax benefits discussed here and through Farm Bill programs or other programs designed to improve conservation on private lands. Nongovernmental aspects of land ownership, like the decision about whether or how to proceed with energy development, also offer both readily perceived benefits and sometimes hidden challenges to the landowners who seek to balance land use with land conservation. Working through these challenges and opportunities with qualified local attorneys will help landowners make the most of their conservation efforts.

CHAPTER 12

Economic Challenges and Opportunities

DREW BENNETT AND CLAY MATHIS

Although chapter 9 touched on financial realities that impact the motivations of land stewards, this chapter takes a deeper dive into the economics of private land stewardship and explores promising emerging markets related to the critical ecosystem services provided to society by well-managed private lands.

Introduction

In 1869, the celebrated conservationist John Muir wrote in his journal, "When we try to pick out anything by itself, we find it hitched to everything else in the Universe" (Muir 1911, 35). As an advocate for public wilderness preservation, he probably was not thinking about stewardship of private lands, so it is striking how closely the words of rancher Pat O'Toole mirror Muir's: "You have to see your operation as part of a complex whole—where everything is integrated and needs to be considered together." Pat O'Toole owned and managed the Ladder Ranch, an operation spanning the Wyoming and Colorado border near Savery, Wyoming, with his wife, Sharon, and their children and grandchildren. He went on to say, "You have to see how all the pieces fit together both ecologically and economically." Although separated by more than a century, Muir and O'Toole both acknowledge that ranches and natural resources are best managed as integrated systems.

The O'Tooles have decades of experience experimenting with different management techniques and a long history of collaboration and partnership with state and federal agencies and conservation organizations. As a result of their efforts, the Ladder Ranch boasts some of the highest densities of Colombian sharp-tailed grouse (*Tympanuchus phasianellus columbianus*) ever recorded and several of the largest greater sage-grouse (*Centrocercus urophasianus*) leks in the region (figure 12.1)—alongside a profitable cattle and sheep ranching business. In recognition of their stewardship, the Sand

FIGURE 12.1. Male greater sage-grouse booming to impress females on the Ladder Ranch (photo credit: iStock Photo / milehightraveler).

County Foundation awarded the Ladder Ranch the Leopold Conservation Award for Wyoming in 2014, yet the O'Tooles are just one example of the many land stewards achieving ecological and financial returns on their operation.

In this chapter, we describe connections between landowner finances and stewardship of private lands, including economic constraints and efforts to align financial incentives with stewardship of land, water, and animals. We also feature the experiences of the Ladder Ranch as an example (though there are many others) to provide context and illustrate points. It is important to note that landowners have diverse motivations for owning and managing land—many of which are not financial (see chapter 7). Some are driven by the hunting, fishing, and other recreational opportunities that land ownership provides. For others, farming and ranching are a lifestyle and the continuation of a heritage that goes back generations. Farming and ranching represent something more fundamental—for a large percentage of landowners they are the basis of a livelihood. Land for these owners often represents their primary business asset with accompanying capital investments in infrastructure, equipment, livestock, and other inputs necessary for running an agricultural enterprise. For these owners all land-based activities must make sense with their bottom line. The pressure to maintain economic viability is great and has generally been a catalyst for diversification of enterprise to spread the fixed cost of land ownership over more revenue streams.

Recognition of the economic value of well-managed private lands is growing—and this value goes well beyond the commodities agricultural operations produce. A wide range of ecosystem services flow from private lands—from water filtration and carbon sequestration to the provision of scenic views and wildlife habitat (see chapter 3). A 2017 study of private lands conserved with conservation easements in Colorado found that these lands contributed \$5.5–\$13.7 billion in ecosystem service benefits to the public—or \$2,700–\$6,600 per acre (Seidl et al. 2017). A similar study of 56 ranches in California estimated that the ecosystem services flowing from those properties provide between \$364 million and \$1.44 billion in public benefits annually (Johnsen et al. 2020). While these are staggering numbers and document the economically valuable contribution of well-stewarded private lands to the public, landowners typically lack ways to monetize these benefits. This results in landowners providing positive externalities (benefits realized by a third party not directly involved in the marketplace) but without a clear price signal to incentivize management in ways that sustain the landowner's business and ensure the ongoing supply of public benefits. There is increasing recognition of the public benefits private lands provide, and efforts are underway to financially support landowners in continuing to provide those benefits. However, economic realities as diverse as global commodity markets and local land-use dynamics place financial constraints on the management decisions of landowners. The following details some of those financial constraints, describes existing efforts to financially support stewardship, and discusses the growing movement to compensate landowners for the ecosystem services their land provides and the inherent challenges involved with this paradigm change.

Economic Realities

Great land managers understand and remain committed to long-term stewardship of water, livestock, wildlife, and people. However, land-based businesses are generally low-margin and variable-margin enterprises. As a result, balancing stewardship with economic goals is a challenge for most landowners. Some investments in land resources have positive short- or medium-term impacts on production enterprise profit, while others have minimal or even negative impacts on the bottom line. In navigating this balance there are several external challenges that create economic constraints on landowners. In many regions, the value of land for agricultural production has diverged significantly from the market value of land as real estate prices are driven up by exurban or rural residential development, amenity buyers, institutional investors, and other dynamics (Gosnell and Abrams 2009; Burns et al. 2018). This divergence of land values from the land's productive potential creates a significant barrier for farmers and ranchers in accessing land to start or expand an agricultural operation and may require taking on substantial debt. The increased cost of land is just one of several expenses to an operation, many of which have increased over the past century relative to the overall prices received for products sold (TLT 2014; Burns et al. 2018). Prices for agricultural commodities, like any commodities sold in competitive markets, can be volatile and fluctuate significantly throughout a growing season and across years. This variability adds uncertainty and risk as producers plan and budget to make their operation financially viable.

The regulatory environment is another area of concern among landowners. Regulation can come in many different forms and from governments at local to federal levels. At the local level, the most relevant type of regulation relates to land-use planning and zoning, which in most jurisdictions is controlled by county governments. Zoning decisions can place restrictions on land-use activities such as limiting the construction of certain types of buildings and their locations or limiting industrial activities to certain areas. Land-use plans and zoning decisions can also impact property taxes if the county assesses the land for its "highest and best use," which may be as a residential subdivision or other use outside of agriculture. In instances where zoning for a property changes from agricultural to other uses, property taxes can also increase dramatically. However, many states and counties provide for a decreased valuation on property taxes if the land stays in agriculture or timber, and some, like Texas, offer a wildlife tax valuation. Increasing land values surrounding agricultural-based operations also serve to increase pressure to sell or subdivide because of increasing land market values.

Federal and state regulations can also impact private landowners. These regulations range from those specific to conserving natural resources to those that impact business operations such as labor employment laws and rules governing how long livestock can be transported. Among environmental laws at the federal level, the US Endangered Species Act (ESA) and the US Clean Water Act are frequent sources of concern for many landowners. Under the ESA, for example, the presence of a listed species or habitat for the species can lead to land-use restrictions that limit how private landowners manage their property if that management results in the significant modification or degradation of the species' habitat (i.e., what would constitute a "take" under the ESA; Baier 2020). Federal efforts like the establishment of Safe Harbor Agreements for ESA-listed species and Candidate Conservation Agreements with Assurances for species under consideration for listing under the ESA can provide some regulatory assurances for landowners that they can continue to manage their properties in similar ways to current practices (Epanchin-Niell and Boyd 2020). Yet the potential for ESA restrictions remains a major concern for many landowners, even if the law impacts only a small number of them (Baier 2020).

Regulations in some situations, like managing labor laws and tax reporting, increase the operating costs of a property. While these issues may seem distant from the topic of land stewardship, they can take significant time and money to manage—leaving less mental energy and resources for land management. In more dramatic instances, regulations can result in substantial restrictions on how a property is managed and may significantly impact an agricultural operation's bottom line. This can have the unintended effect of pitting landowners against wildlife and potentially even result in contrary incentives to landowners whereby habitat for a listed or candidate species is a liability rather than an asset. This dynamic played out on private forests in the range of the red-cockaded woodpecker (*Leuconotopicus borealis*) in the southeastern United States. Habitat for the species required older pine forests, preferably longleaf pine, suitable for creating nesting cavities. Concern over potential ESA restrictions if red-cockaded woodpeckers were discovered on their properties led some landowners to plant smaller, faster-growing pine species and harvest trees prior to them being mature enough to support woodpecker colonies (Byl 2019; Baier 2020). While hosting habitat for endangered species is something that should be celebrated and incentivized, the implementation of regulations like the ESA have resulted in unintended consequences and perverse outcomes for listed species and other environmental concerns. The collective impact of the regulatory environment is of significant concern for many landowners, something they monitor regularly, and can raise uncertainty about operations going forward.

Financial Assistance for Stewardship

A wide range of programs exist and new ones are emerging to support landowners in stewarding their properties (Bennett et al. 2018). Many of these programs are offered by federal and state natural resource agencies and can take the form of cost shares, incentive payments, rental fees, or conservation easements. Cost-share programs split the cost of implementing a specific conservation action or installing conservation-minded infrastructure, such as installing "wildlife-friendly" fences that allow species like deer and pronghorn easier passage. Often an agency will cover 50% or more of the costs, with the landowner covering the remainder. With incentive payments, landowners receive a direct payment for taking a specific conservation action, and payments are often tied to the number of acres (or other relevant metric) impacted. Other programs pay a rental fee for a specific amount of time for a conservation practice or to avoid or limit agricultural production in a specific area for a period of time. Conservation easements are voluntary legal agreements that landowners enter with a government agency or qualified not-for-profit land trust that limits subdivision and development potential of a property. In exchange, landowners may qualify for tax benefits such as federal tax deductions or in certain instances may be able to receive a onetime payment for a portion of the conservation easement's value through a "bargain sale," in which a landowner receives cash for a portion of the conservation easement value and donates a portion (e.g., receiving 75% of the value in cash and donating 25%).

With incentive payments, rental fees, and conservation easements, landowners typically receive direct financial compensation that they can invest back into their operation or otherwise spend how they choose. With cost-share programs, landowners do not receive direct compensation but may be able to offset infrastructure improvements or cover a part of the costs associated with management practices that benefit the operation or meet other property goals. We discuss specific examples of these types of mechanisms later but recommend Baier (2020) for a thorough description of the range of conservation programs available to private landowners. We note, however, that specific programs—whether federal, state, or private—change over time with new programs being added and existing programs being merged or eliminated. Many of the same characteristics of the programs described here are likely to continue in some form or under a new name (and accompanying acronym), but we recommend understanding the big picture of what these programs aim to accomplish and how they are implemented rather than getting hung up on the specifics of any individual program.

At the federal level, the US Department of Agriculture's Natural Resources Conservation Service (NRCS) is the largest agency with a primary mission of working on private land stewardship issues. Operating in all 50 states and US territories, NRCS offers an alphabet soup of programs funded by the US Farm Bill, colloquially referred to as Farm Bill programs, to address a wide range of environmental and natural resource issues. Among the most popular are the Environmental Quality Incentives Program (EQIP) and the Conservation Stewardship Program (CSP). Both programs provide a great deal of flexibility on the specific resource issues addressed and specific practices landowners can implement. EQIP can take the form of cost shares or incentive payments depending on the practices implemented, and contracts typically last 5 to 10 years. Many wildlife-oriented practices are included in EQIP, which is the primary program for implementing species-focused efforts like the Sage Grouse Initiative and the Lesser Prairie Chicken Initiative (both initiatives now part of NRCS's nationwide Working Lands for Wildlife), among others (Naugle et al. 2020). CSP has many similarities to EQIP, but a participating landowner must enroll all eligible acres into the program, creating a "whole farm" program (Baier 2020).

Partners for Fish and Wildlife (often simply referred to as the "Partners Program") is the US Fish and Wildlife Service's (USFWS) primary program for voluntary conservation on private lands. The Partners Program provides technical assistance to landowners in implementing Farm Bill programs and its own cost-share and incentive payments for stewardship projects. Through the Partners Program, and in collaboration with their local conservation district, NRCS, and the Wyoming Game and Fish Department, the O'Tooles constructed 42 structures to improve instream habitat and facilitate fish passage along several miles of Battle Creek (figure 12.2). Several of these structures also integrate improved outdated irrigation infrastructure, resulting in a balance between agricultural needs and stewardship of the property's water resources.

The Conservation Reserve Program, administered by the Farm Service Agency (FSA), is the largest land rental program in the United States. Originally developed to temporarily take highly erodible land out of agricultural production, it has since evolved to also address issues like water quality and wildlife habitat. Participating landowners enter 10- or 15-year contracts and agree to remove cropland from production during the contract period and revegetate the land with native grasses, cover crops, or other vegetation to conserve soil and improve habitat. A subset of the CRP program known as CRP Grasslands, now allows rangelands to be enrolled, which expands the program from the historic focus on cropped lands and allows landowners to receive a rental payment for conservation-oriented management of grazing lands.

NRCS is also a major funding source for the purchase or bargain sale of conservation easements through the Agricultural Conservation Easement

FIGURE 12.2. Instream structures on the Ladder Ranch installed to improve habitat and fish passage along several miles of Battle Creek (photo courtesy of Drew Bennett).

Program–Agricultural Land Easements (ACEPALE). The conservation easement must first be appraised by a qualified real estate appraiser to determine the value of the conservation easement. The appraiser must first determine the value of the property without the conservation easement—known as the "before value"—which is typically done by looking at comparable sales of other properties in the vicinity. The appraiser will then provide an opinion on what the likely value of the property will be once the conservation easement is placed on the property—known as the "after value." This is also typically done by analyzing comparable sales but is much more difficult since only a fraction of properties sold have conservation easements relative to the real estate market as a whole. Since the conservation easement typically restricts subdivision and development of the property (and potentially other uses), the property is hypothetically worth less after the conservation easement is in place. The difference between the before and after values is the value of the conservation easement. The NRCS ACEPALE program currently pays up to 50% of the appraised value of the conservation easement, or 75% if deemed a grassland of special significance, and the remaining value must be matched by other nonfederal grants or donated by the landowner. Many states also have their own sources of funding for conservation easements (see TPL 2019 for a summary of funding sources in each state).

Selling a conservation easement can be part of a financial strategy for agricultural operations. Although landowners often donate a portion of the easement value when doing a bargain sale, the cash landowners receive can be reinvested back into the operation, used to fund retirement, or spent any other way a landowner wishes. This strategy can help address the farmer and rancher issue of being "land rich and cash poor" by being able to extract some of the real estate value out of the property while still owning and managing the property for agricultural production. The O'Tooles used this strategy and reinvested proceeds they received from placing part of their property under conservation easement to purchase a neighboring property that included the ranch's original homestead and expand their operation. While incorporating conservation easements as part of a financial strategy may be viable for the current generation, once the land is under easement, future generations likely will not have the same opportunity available to them. Landowners should understand and fully consider the long-term impact of a conservation easement on land asset value in the next generation.

In summary, although there are a number of federal and state programs to economically support landowners with stewardship, these programs have their limits. These programs do not address the broader market forces impacting agriculture and rural land use in the United States. While landowners can tap into programs for modest revenues or cost shares to implement infrastructure or other practices that are compatible or enhance productive uses, these are typically insufficient to fully support an economically viable agricultural operation and address trends in land conversion, landscape fragmentation, species decline, and other environmental concerns. Given these limitations, there is a growing recognition that a paradigm change is needed to support land stewardship by recognizing and rewarding the value of stewardship.

A Paradigm Change Recognizing the Value of Stewardship?

There is increasing recognition that the management and input cost of stewardship is largely supported by the landowner. Whether it is addressing invasive species through spraying or other control methods or maintaining wildlife habitat for nongame species, landowners have direct costs associated with management that may reduce revenues. Recognizing the multiple public benefits derived from private lands, there is a gap between the economic values of those benefits relative to the monetary value landowners can generate from their land. New approaches and opportunities are emerging that represent a paradigm shift in supporting stewardship on private lands and helping to close this financial gap.

A new vision of the "ecosystem service commodities" private lands can produce is emerging that builds on the existing products agricultural operations can monetize—such as livestock, timber, crops, and hay. These new commodities include carbon sequestration in forests and soils, water filtration, flood attenuation, habitat for species of concern, and recreation. There are emerging models and pilot efforts to generate revenue streams from this broader portfolio of ecosystem service commodities, several of which we highlight in the following discussion (figure 12.3).

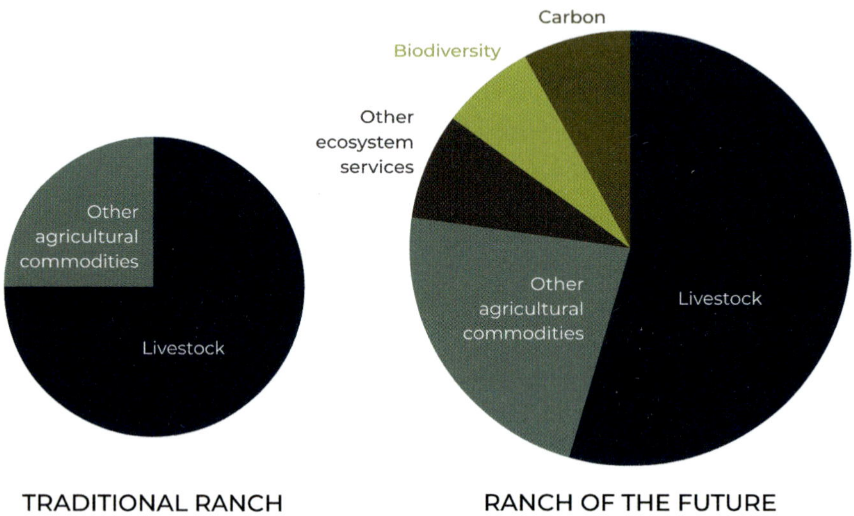

FIGURE 12.3. A hypothetical example of revenue sources from a "traditional ranch" and a "ranch of the future." Selling new "ecosystem service commodities" can diversify income streams and grow revenues while ideally having no or minimal impact on revenues from traditional agricultural products like livestock, hay, timber, and crops.

Carbon markets continue to hold promise as a source of income and a way for landowners to monetize an ecosystem service. While the Chicago Climate Exchange and other efforts in the early 2000s largely became moribund around 2011 (Knight et al. 2011), natural climate solutions are likely to be an important component of a comprehensive strategy to address climate change (Fargione et al. 2018). In simple terms, the carbon market can be divided into compliant (i.e., regulatory) or voluntary markets. In the current compliant market, the most relevant opportunities available to landowners are forestry projects with credits sold through California's Air and Resources Board's Cap-and-Trade Program. Through this program, landowners can generate carbon credits by adopting forest management practices that increase aboveground biomass, typically by shifting timber harvests to longer rotations or conducting selective harvests that maintain forest cover. Current forest carbon markets are likely economically viable for private forest owners only in regions with rapidly growing forests like the Pacific Northwest and the Northwoods of Minnesota, Wisconsin, and Maine. Outside of California's compliant market, voluntary markets also exist for similar forestry projects and for private grassland or rangeland projects that prevent conversion to other land uses, like crop production, that result in the release of carbon sequestered in the soil. In these "avoided conversion" projects, landowners can develop credits for certain soil types at risk of conversion by committing to not converting their land through a term (e.g., 100 years) or perpetual easement.

Although opportunities exist for landowners to participate and benefit from the carbon market, the current price of carbon (~$10/**metric tons of carbon dioxide** [mtCO2] on the voluntary market) limits the financial viability for many landowners, especially given the high transaction costs required to access the market. Other opportunities to compensate landowners for carbon sequestration beyond true market approaches are also emerging and show promise. The company Bayer operates a program to compensate farmers for "climate-smart" agricultural practices. Enrolled participants receive $3 per acre for no-till/strip-till practices, $6 per acre for planting cover crops, or $9 per acre for both practices. While modest in terms of financial compensation, it provides an example of the growing commitment many corporations are making toward climate responsibility that can create opportunities for landowners. In a similar vein, the nonprofit Ecosystem Services Market Consortium is working with over 60 partners from the corporate, nonprofit, and agricultural sectors to develop markets and incentives for agricultural producers to reduce climate impacts and enhance the sustainability of supply chains (Buckley Biggs et al. 2021).

A number of other markets have been established to conserve water quality, wildlife, and other resources. In Oregon's Rogue River basin, the City of Medford is leasing land from private landowners along the river to restore the riparian areas and generate temperature credits due to the cooling effect riparian trees provide by shading the river. Medford uses the credits to offset the temperature impacts of effluent from its wastewater treatment facility and remain in compliance with the Clean Water Act. Landowners in the watersheds that are the source of drinking water for cities like New York City; Eugene, Oregon; and Tulsa, Oklahoma, can receive financial incentives for adopting land management practices that filter the water or reduce erosion and other ecosystem services (Bennett et al. 2014). Markets have also been developed to provide incentives for landowners for conserving habitat for species of concern—including ESA-listed species. For example, the State of Nevada established the Nevada Conservation Credit System to allow landowners to generate credits for habitat improvements for the greater sage-grouse on their properties. These credits can then be sold to other entities, like energy or mining companies, that need to offset impacts to sage-grouse habitat required by state regulations. Similar habitat markets have been established for the monarch butterfly (*Danaus plexippus*), dunes sagebrush lizard (*Sceloporus arenicolus*), lesser prairie-chicken (*Tympanuchus pallidicinctus*), and several other species (Kreuter et al. 2017). Outside of habitat mitigation markets, a number of stakeholders are developing the concept of a "habitat lease" that would directly compensate landowners for providing important habitat for target species in Wyoming and other states (WLA 2021).

These ecosystem services approaches, while not widespread, have increased substantially over the past couple of decades and point toward the growing recognition of the many benefits private lands provide. While not a silver bullet, these approaches represent a step toward better aligning economic incentives with stewardship. Despite this progress, however, market-based approaches to stewardship face substantial obstacles in becoming mainstream and broadly accessible to landowners. Many programs remain in pilot stages, and their long-term viability is not ensured. Other programs have a limited geographic scope, such as a specific watershed, and participation is limited to landowners in those specific locations. Some market-based approaches are also limited to landowners with specific resources, like high-quality habitat for a species of concern, which may limit the number landowners who can access the market. Many market-based opportunities are also driven by regulations through which landowners develop mitigation opportunities for entities that negatively impact a resource. Regulatorily driven programs can falter when there is uncertainty over the regulations. For instance, several habitat credit trading programs were developed ahead of a listing decision for the greater sage-grouse but were largely unnecessary after the USFWS decided not to list the species for protection under the ESA in 2015—essentially eliminating demand for credits, as entities would not be required to mitigate habitat impacts under federal law. Market-based approaches to incentivize or reward stewardship are often complex with technical evaluations that need to be completed to calculate credits generated by landowners or that need to be offset by developers (Bennett and Gosnell 2015; Buckley Biggs et al. 2021). The technical basis for the credits can be upended if new science challenges assumptions in the crediting process. Collectively, these challenges present a significant barrier to the paradigm shift of more fully compensating landowners for good stewardship and the provision of ecosystem services.

Conclusion: "The Custodian of Two Interests"

Writing in 1934, Aldo Leopold recognized a fundamental challenge of private land stewardship: "The crux of the problem is that every landowner is the custodian of two interests, not always identical, the public interest and his own" (cited in Meine and Knight 2006, 160). Although ecologists and economists have made significant advances in understanding and quantifying the ecosystem services private lands provide, Leopold's insight is no less relevant today as stewards of private lands navigate the rights and responsibilities of land ownership. Solutions to balancing this dual nature of land management range from the cost-share and incentive programs discussed previously to efforts to compensate landowners for new ecosystem service commodities. Each approach brings its own ecological and economic benefits—and limitations. As if anticipating contemporary debates, Leopold continued:

> What we need is a positive inducement or reward for the landowner who respects both interests in his actual land-practice. All conservation problems—erosion, forestry, game, wildflowers, landscapes, or what not—ultimately boil down to this. What should this reward or inducement be? What is a practical vehicle for it? These are the two basic questions in American conservation. (160–61)

Conservation practitioners, government programs, and landowners continue to grapple with these fundamental questions nearly 90 years after Leopold penned this essay. Yet there is cause for optimism. The range of innovation and experimentation in developing positive inducements or rewards for private landowners is greater than in any other period in American history. Some of these innovations and experiments have failed and will fail. A number have shown considerable ecological impact and are broadly popular among landowners, while others hold considerable promise.

The economics of production agriculture will continue to present landowners and managers with significant challenges for the foreseeable future—many of which span well beyond the domain of land stewardship and conservation. Bridging the gap between the economic costs of stewardship that benefit public interests with the financial returns of agriculture and other private enterprise can assist landowners in resolving the fundamental crux that Leopold identified. For Pat O'Toole, experimentation has been key to developing a ranching system that includes economically viable livestock enterprises and benefits the wildlife and natural resources they steward. "Part of our success can be attributed to the fact that we're willing to make lots of small mistakes in order to have big successes." Ultimately, all landowners will need to forge their own path and develop solutions that work for their operation and their land's ecology. There are likely no panaceas, but there are more tools in the toolbox than ever to assist in this effort. Even more important, the stories of the O'Tooles and many others demonstrate that there are innumerable inspiring examples of landowners across the United States addressing the problem head-on with their own mix of practices, innovations, and commitment to the land.

SECTION V

Fostering the Stewardship Pathway

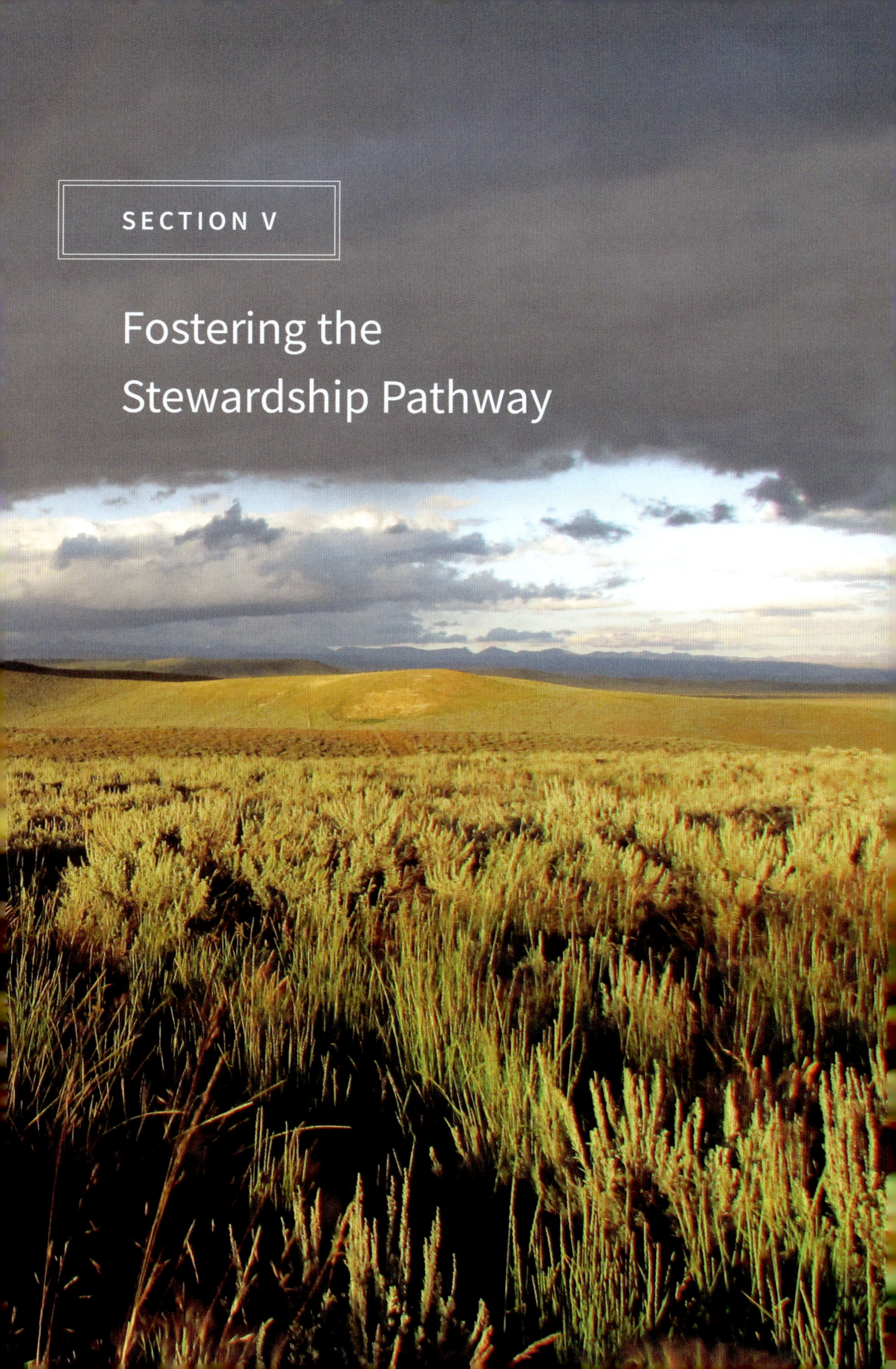

CHAPTER 13

Land Stewardship

Agency Perspective

WENDELL GILGERT WITH KEN CEARLEY

With decades of combined experience working as agency practitioners with land stewards from California to Kansas, the authors of this chapter share lessons learned and best practices for interacting with private land stewards to accomplish shared goals for natural resources. Their timeless insights will help guide young professionals as they navigate new relationships and competing interests in a way that meets their organization's goals while serving as a valuable resource for private land stewards.

Introduction

Engaging working land stewards (farmers, ranchers, foresters, miners, private consultants, etc.) combines art, science, spirituality, economics, sociology, culture, emotion, and sometimes even politics. Most landowners think of themselves as stewards or caretakers of natural resources, domestic and wild animals, and other land-based products or benefits. By definition, stewardship is the act of conducting, supervising, and managing through careful and responsible actions something entrusted to one's care. Stewards are, above all else, passionate about their responsibility to their crops, livestock, trees, or simply stated, the *land*. They are known to take any possible measures to shelter, guard, or otherwise protect the plants and animals in their charge from weather, disease, fire, or other natural occurrences and even their fellow human beings when warranted. In this chapter, we provide a brief profile of the land steward based on a century of collective professional experience. More important, *especially for those who are just now or have recently embarked on a career in conservation and restoration*, we offer a sample of considerations and approaches to cultivate, enhance, encourage, grow, and support land stewards.

Over the decades and with the help of innumerable colleagues, farmers, ranchers, foresters, miners, and private consultants, we have developed an outline of land steward characteristics. As you work with land stewards in

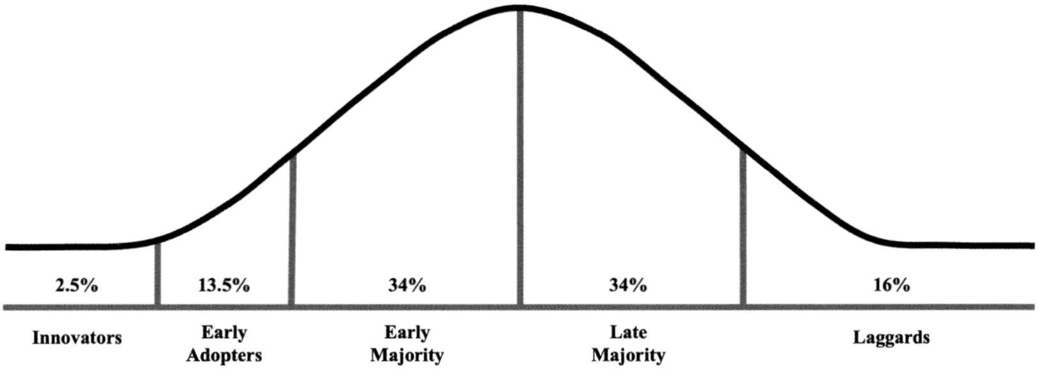

FIGURE 13.1. Theoretical land stewardship adoption curve (modified from the product adoption curve popularized by Everett Rogers in his 1962 book, *Diffusion of Innovations*).

your career, you will certainly encounter individuals on all points of the sociological curve of adopting land stewardship (figure 13.1). This adoption curve, commonly used when designing marketing strategies, can be a valuable concept to consider when evaluating landowners and where they may be in their stewardship journey. By understanding a particular landowner's relative position on this curve, a natural resource practitioner is best able to meet them at their level and provide the most appropriate resources, tools, and support to get them where they want to go.

As conservationists, we find working with innovators and early adopters to be relatively easy and satisfying. Your real challenge will be working with the late majority and the laggards—or those right at the tail-end of the curve. This group may be challenging to engage because progress can be slow and at times frustrating; however, when headway is made with them toward focused land stewardship, it can be extremely satisfying.

Who Are Land Stewards?

As Aldo Leopold wrote, "The landscape of any farm is the owner's portrait of himself" (1991, 263). That landscape signature is usually very visible to both the community and visitors to the land. While many individuals and groups consider themselves direct or indirect stewards (agency personnel, land trusts, volunteers, environmentalists), this chapter is focused on the direct stewards of our private lands—the landowners, managers, lessees, and private consultants who make day-to-day decisions about the health and management of the land in their care. Following is a discussion of defining characteristics of land stewards.

Desire to Improve Land

Land stewards provide service to the land, making daily decisions where they must choose between what is best for the land and what is economically feasible for their operation. For example, most grain and hay producers must make decisions based on the need to manipulate (e.g., weed, harvest, mow) their crop in a timely manner, while also accounting for nesting and fawning/calving wildlife who use their fields. Will they postpone harvest, risking lower quality or yield of their crop, which can lead to lower prices? Do they utilize techniques (e.g., flush bars, mowing inside out, no nighttime mowing or harvesting, slowing the speed of their equipment) to minimize wildlife injury or fatalities? Or do they charge ahead with little thought to the consequence to their wildlife resources? Most land stewards will sacrifice some economic benefit for the greater good of their fish and wildlife resources, but there is a point at which they must do what is economically practical to pay bills and stay on the land. Feelings toward the land and the resources it supports run deep and fuel land stewards' passion for their activities. As you grow to know land stewards and their operations, you will experience a spectrum of land steward emotions. Some are very emotional, showing excitement over the smallest change—new plant species, newborn lambs, or new bird sightings—while others are quietly stoic and reserved with their feelings and actions. However, as you gain their trust and get to know them more deeply, those emotions will often rise to the surface.

Curious Minds

Land stewards do not have idle minds but instead are curious, inquisitive, and deeply interested lifelong learners. Land stewards seem to embrace new technology, management techniques, and novel or innovative ideas related to their enterprises and the land. Moreover, you will find as you work closely with land stewards that they have a wide variety of interests. It is in the individuality of those interests that you will find the varying arrays of their individual preferences and approaches. Like Aldo Leopold and other keen observers, land stewards put into action what they learn from their observations. The ability to share those observations, and then take the time to reflect on the outcomes, is a hallmark of a land steward. The process of observing, participating, and reflecting means that land stewards serve as the innovators and early adopters that are usually found on the vanguard of the development and experimentation of new management, conservation, or restoration techniques. Obstacles that become challenges often find solutions when they are shared widely among land stewards in their communities and beyond.

CASE STUDY 13.1

Brian Treadwell

Brian Treadwell is a sheep and goat rancher in West Texas and credits a couple of key agency practitioners for the development of a generational land ethic in his family. These practitioners cultivated relationships with area landowners, leveraging their sense of community and desire to learn from one another to encourage large-scale use of prescribed fire in areas where it had been historically underutilized.

"Our state wildlife biologists pulled landowners together and took us to wildlife management areas so we could see, firsthand, how our country could look with the use of prescribed fire. It was hard to deny that fire would work on our country when you could see with your own eyes what is going on next door."

Deep Understanding

Land stewards have a detailed understanding that their land, natural resources, natural systems, and the agrarian operations are a fabric of interconnected natural, economic, cultural, political, educational, and spiritual planes. Pull one string of that quilt and it is tied to all others. Being in tune with the land means stewards must become land, using all senses to diagnose land health by assessing natural processes (water and nutrient cycles, succession, energy flow, food pyramids, pollination, photosynthesis) and where land health improvements are needed and then prescribe a range of conservation practices to improve the conditions on the land.

Having a holistic view, stewards understand they are caretakers of the land for the duration of their lives and hopefully, for future generations. Most strive to leave their land and resources in better condition than they found it. They ascribe in a similar way to the Seventh Generation Principle, an ancient Haudenosaunee (Iroquois) philosophy, in which the decisions we make today should result in a regenerative world at least seven generations into the future.

Since most land stewards have chosen their agrarian lifestyle, they gain an intimacy with the land and come to know and understand that there are a rhythm and pace to conservation, restoration, and stewardship—that the land did not develop into its current condition overnight, and restoration and conservation are long term and incremental. It takes patience to see that land management actions mature, and it takes presence of mind to ponder the outcomes, adjust, or stay the course. They recognize and understand that they will need much help from family, neighbors, friends, colleagues, and agrarian professionals along the path to maintain or restore land health.

CASE STUDY 13.2

Jim Hebbe

Jim Hebbe and his father were early adopters of no-till practices on their east-central Wisconsin grain farm in the early 1980s. Jim describes how mistakes in farming are what led to learning experiences and development of more sustainable practices. Common practice when Jim was a teenager was to till the ground in the fall because the belief was that by creating greater surface area, the soil temperatures would rise sooner in the spring and allow for earlier planting. Jim says that "a turning point in my life was one year when we had tilled in the fall, and by mid-March the winter runoff was particularly heavy. We watched as the topsoil of our farm was carried downstream, and my dad said, 'We're never going to do that again.' It would have been easy to say, 'Well, that was an unfortunate act of God that we can't do anything about,' but my dad was accountable and instead said, 'This is a problem we created, and we're going to have to be the ones to fix it so it doesn't happen again.' That was the beginning of no-till for us."

FIGURE 13.2. Jim Hebbe, recipient of the Leopold Conservation Award for Hebbe Farm in Green Lake, Wisconsin (photo courtesy of Jim Hebbe).

Experiential Learners

Life for a farmer or rancher can be among the riskiest of occupations, given extreme weather events, disease, weed or insect infestations, predators, and volatile markets, all of which require that a producer be skillful in decision-making. Those decisions are not always successful, which means land stewards must be resilient to move away from failures and adapt to new directions with the information they learned. Successful land stewards learn from their mistakes. Their successes are not only remembered and repeated but often shared with neighbors and friends to help them prevent the same mistakes. The best conservation, enhancement, and restoration actions are often well thought out, included in the conservation plan, and enthusiastically managed following implementation. Whether they are shifting rangeland pastures to include more native grass species, prescriptively grazing riparian zones to increase structure for wildlife, integrating a cover crop, planting a hedgerow or shelterbelt, or aggressively taking measures to reduce sediment in their streams or waterways, the continued improvement in overall land health is a hallmark of land stewardship.

Helpful to Others

Farmers, ranchers, and foresters, no matter their political, cultural, ethnic, or social persuasion, are often known for their positive interactions with their neighboring working land stewards and community. Whether it is lending equipment, helping work livestock, assisting with harvest, offering their land for workshops, or sharing their innovations, land stewards are often first to volunteer. Most rural community fire departments, school boards, churches, and youth activities (4-H, FFA, Little League, Scouts, etc.) rely heavily on local volunteers.

Do not expect most land stewards to be loud or boastful about decisions or accomplishments related to their production, quality of life, landscape, conservation, or restoration goals. With gained trust, you can assist with the development of their various resource-specific plans (e.g., ranch, farm, animal waste, irrigation water management, nutrient management, timber harvest, silvicultural, grazing management). During those planning processes, you will be consulted on numerous decisions related to the inclusion of various conservation, restoration, and management practices. Consideration of these can take several days, weeks, months, or even years to reach a decision about which direction they choose to pursue. Even then, when a decision about management direction or conservation/restoration practice is decided, needed regulatory permits, weather, and project financing can further delay implementation. You will need to exercise patience and understanding during this interval.

Increasingly, land stewards are trending away from conventional agriculture that relies on a variety of artificial fertilizers, herbicides, pesticides, or high-cost precision technology in favor of pre–*Silent Spring* regenerative agriculture that embraces natural fertilizers (manures) and pesticides, as well as the wide selection of regenerative conservation practices such as cover crops, crop and livestock rotations, use of manures, hedgerows, shelterbelts, and other conservation buffers (e.g., field borders, grassed waterways, riparian forest buffers, and filter strips). Further, fish and wildlife are integral components of their everyday life. They understand much of the life history requirements of various fish and wildlife species that occupy their and neighboring land. Thus, if critical loafing, nesting, escape, or thermal cover is not available, they will attempt to provide those missing components or structures (nest boxes, brush piles, planted habitat) and manage their resources and land to the best of their ability and knowledge.

Perhaps the trend is fostered by the generosity and sharing of other land stewards' understanding, perceptions, successes, and failures. When new, inexperienced, or transitioning land stewards begin farming, ranching, or tending to their forestland, their neighbors traditionally will offer advice, equipment, or even their own labor to aid their new neighbors and teach new skills. In any occupation, having a mentor can be the difference between a successful or unsuccessful career, and this is particularly true with working lands. There are numerous elements related to every aspect of working the land, be it livestock breed performance and behavior, efficacy of various types of equipment, timing of various crop cultural practices, or transitions from conventional to regenerative farming or ranching. Having guidance, support, and a level of security from a producer who is a veteran of navigating management of those working lands is invaluable.

Techniques to Engage, Enhance, Encourage, and Cultivate Land Stewards

This section is intended to convey decades of experience working with land stewards and is intended for you to benefit and learn from our uncountable mistakes and failures. More important, we are offering the positive and hopefully useful lessons we have learned and experienced working closely with an amazing variety of land stewards over our careers. Early in our careers, we have a deep desire to prove our mettle and worth, but it is important to avoid the tendency to impress land stewards with your newly acquired knowledge, much of which came from academic textbooks, classes, or published research. When presenting novel ideas to producers, it can take weeks, months, or even years for them to see value or a place in their operation where that idea or concept fits or is of value to them. Keep in mind that delays in accepting your ideas may be more likely attributed to the time it takes to build trust than the relative validity of your advice.

First Impressions

When building trusting relationships, you only get one chance to make a first impression. Before embarking into the field, review the following checklist:

- Dress appropriately based on your audience, while still being true to yourself. Be careful not to underdress to the point that they might think you have a low-standards impression of them and are therefore trying to dress as you think they would.
- Always give (if culturally appropriate) a firm handshake to show you are serious and have confidence.
- Have a good sense of humor and display genuine evidence of it. Smile and be lighthearted, even self-deprecating if appropriate at the moment.
- Show quiet confidence in your demeanor—considering before responding, thinking before abruptly and definitively giving an answer.
- Make sure your comments and suggestions are tailored to their particular situation and challenges, acknowledging that they have the ultimate say and will make a decision based on a suite of factors, some of which you might not be privy to.
- Listen, rather than talk! Ask questions but be brief; plan for 20% talking and 80% listening.
- Meet on their terms/time lines, which may be after hours or on weekends to accommodate their schedule.
- If you do not know the answer, tell them and offer to find the answer if possible. You will gain more respect by saying you do not know than by making up an answer that may not be correct.
- Be humble—landowners are generally offended by government/conservation workers with strong egos. Always be aware of what you say that may create a perception of being boastful, pretentious, or dominating.
- Be passionate about landowners and their project or potential project, not about your opinion on the subject.

Once you have made a good first impression, it is time to start building the foundation of a strong working relationship. As mentioned in the checklist, plan to ask meaningful open-ended questions that do not elicit simple yes or no answers. Questions using who? what? when? why? where? and how? usually require detailed answers, which will frequently provide the information needed to fill knowledge gaps for resource inventories, management approaches, monitoring, or intended conservation or restoration results.

There are few activities that gain as much traction with a producer than spending time walking in their shoes. That can mean spending early mornings, late evenings, or weekends to accomplish that objective. Ask to help in ways that are useful to them and where you are becoming more intimate with their operation(s). Lend a hand if you have opportunity, as there are few things more off-putting to a producer than upon your arrival for a field visit, they are involved with a chore or task, and you just stand around waiting for them. Always ask if you can assist. Both you and they

CASE STUDY 13.3

Jimmy Rutledge

Jimmy Rutledge served as a private lands and technical guidance biologist for the Texas Parks and Wildlife Department for 30 years. Jimmy describes the evolution of his career as one of both learning from landowners and learning how to meet them where they are to help foster the development of their individual land ethic. "I have always been passionate about plants, and especially using an understanding of plant species diversity as a gauge of ecosystem health. I learned quickly that most of the landowners I worked with in South Texas, at least initially, had very little interest in plants or ecosystem diversity. What they had extreme interest in, however, was growing big white-tailed deer. I eventually learned that I could use their interest in growing big deer to cultivate an understanding of and appreciation for the importance of ecosystem diversity. By pointing out important deer browse plants, it gave me the opportunity to talk about the larger community of plants and how those plants contributed to deer health and antler development, but, 'Look—there's a Bell's vireo! And look over here at these monarch butterflies on this flowering forb!' In the end, I found that while white-tailed deer or bobwhite quail were what got me in the gate initially, many times it was the ancillary ecosystem benefits that would have landowners calling me back."

FIGURE 13.3. Jimmy Rutledge, retired Texas Parks and Wildlife Department biologist (photo courtesy of Jimmy Rutledge).

know you may not have the skill or knowledge about the task, but at least ask. You will be surprised how often you can help, and the producer will not forget those gestures. Also, offers of help when you know the producer is shorthanded will gain even more focused attention from the producer when time allows. Remember that farms, ranches, and forests care nothing about our calendars and do not recognize a 40-hour workweek. Land stewards must respond to whatever comes their way in terms of weather, diseases, a bloated cow or ewe, a breached calf or lamb, equipment breakdown, broken pipelines, and so on. Keep in mind that the vagaries of everyday land management continually carry surprises, requiring you to respond in a positive and productive manner when help is needed or meeting times quickly change.

Nothing initiates open dialogue more than connecting with the interest of the producers, so pay attention to key features of their operation. As you enter a property for the first time, it is smart to note the hobbies, interests, or patterns of the owner/operator, like types of crops or livestock, equipment brands, fence types, landscaping, and even hobbies (e.g., building bird houses, collections of old farm equipment, Native American artifact collections). Noticing and positively commenting on their interests will usually open conversation and lead to greater trust with the producer.

Paying attention in this way can be extended from observations to common vocabulary and language, so spend time prior to your visit learning specifics about local geography, crops, livestock breeds and classes, and forestry trees. Each has their own vocabulary. For example, sexual designations for numerous kinds of livestock (e.g., gilt/sow, heifer/cow, ram/wether, bull/steer, or stallion/gelding). Likewise, there are tremendous varieties of cultural practices that require specific equipment, whether it be logging, farming, or ranching. Techniques using and managing diverse kinds of equipment, crops, and livestock and trees can vary regionally and even from property to property, so it is beneficial to find mentors to help navigate local language and customs.

As mentioned earlier, land stewards are often the innovators and early adopters of new ideas, approaches, or concepts in their field. When you see or hear of someone employing innovative or useful management or restoration techniques with success, give, without going overboard, a well-placed compliment that will be forever appreciated, and that action will permeate through the community, yielding multiple benefits. Site visits and interactions with land stewards are steps to becoming part of the local community, and they come with responsibilities. *A career in natural resources conservation or restoration is NOT a nine-to-five occupation.* While you are most often paid a salary for a set number of hours per week, you will find that if you want to become established and make a difference in your community, you may be called on before and after your workdays, evenings, and especially on weekends—times when the land stewards are not completely occupied with managing their operations. As you become a familiar resource and face in your community, there will be ample opportunities to assist shorthanded land stewards with working their livestock, or assisting with planting or harvest or with community educational activities, like workshops, field days, 4-H or FFA activities, and local and county fairs. All are great opportunities to build relationship capital and credibility in your community with the goal of becoming a highly regarded community asset or organizer.

These techniques are not meant to be a facade, so always be yourself and do not try to change who you are to "fit in." There are some obvious limits to this suggestion, and you should avoid statements or politically controversial T-shirts, hats, or other advertisements that show your political views. As with most occupations, it is best to stay away from the traditional controversial topics of politics and religion.

Relationship Building

Having a relationship means that interactions will occur between or among individuals, so deliberately ask to learn from your farmer, rancher, forester, or miner cooperators. Here are a few items that could prove very beneficial to you:

- There is a pace to working with land stewards that is governed by your ability to build trust and learn local geography, culture, and customs. Community-based conservation is like a marathon, not a sprint.
- You are never too old to learn something new, and all perspectives on an issue are important. Listen, listen, listen.
- There is a reason some landowners do not like agencies and conservation groups. Do not be offended by that but instead figure out where we went wrong and try to use your credibility to fix the problem.
- Ask landowners what they need instead of starting off with what you want to deliver. If they need or want something that you cannot deliver, find someone who can and bring that person/group into the conversation.

- Show up at weed pulls, spray days, evening meetings, workshops, and so on. Landowners notice your work ethic and efforts to integrate into their community.
- Host workshops or field trips to build a sense of community.
- Answer questions honestly. Be willing to consider that you might be wrong. If you are so sure that you are right that you are not willing to listen to the other side of the story, you are *probably* wrong.
- Vent with a spouse or trusted friend, but never to landowners.
- Be adaptable—no matter how far down the wrong road you have gone, there is still time to turn back. Let's face it; we have had some success in conservation but have also had a lot of failures. Do not be afraid to change the way you do business.
- It is important to monitor the words you speak, as well as keeping your word, because this defines who you are and what people think of you.

CASE STUDY 13.4

Ken Cearley

Ken Cearley, a rancher, wildlife biologist, private ranch and wildlife management consultant, and retired wildlife program specialist with Texas A&M AgriLife Extension Service talks about the importance of relationship building in garnering trust from landowners: "At the beginning, many years before I joined Extension, in my passion to 'spread the word,' I thought [landowners] would fall all over me because of all the knowledge I acquired from college. In those early days I didn't fully understand just how many factors would go into their decisions. That was frustrating at first, but I finally realized that I just needed to be patient and not take it personally if they didn't accept my recommendations right away, or ever."

FIGURE 13.4. Ken Cearley with rancher Johnnie Hudman (photo courtesy of Jenny Sanders).

As you gain experience with land stewards and build a résumé outside of the university, thinking of ways to convey your experiences often leads to the opportunity to tell a good story. Be ethical and stay away from embellishments. Few of us have the storytelling skill of someone like Aldo Leopold or Walt Whitman, but we each have a unique voice and perspective. Use your experience, viewpoint, and voice to craft your unique story, but remember to "stay in your lane!" In the natural resource's world, it is useful to employ a holistic view, where all resources are connected. Because the physical and biological natural sciences are so closely related, most of us know at least a little about many aspects of natural resources. Avoid taking what you think might be true or possible and representing it as absolute truth. There are few actions that will diminish your credibility with land stewards more than representing something as fact when it is not.

Building Your Network

To be a successful natural resource professional, you must be willing to "go the extra mile." Many times, this means bringing other partners along in the effort in order to cultivate resources and ensure a multidisciplinary approach. Consider these items as you work to build your conservation network:

- Sound science must guide us in everything we do. If we are not working in key areas, then we are not doing the right thing for the resources and the American public.
- Once you have the science to identify where best to work, and what practices are most appropriate, then it is time to engage community members and partners.
- In our profession, we need to figure out how to help or lead from behind. Being out front may not always be the most productive position. When you can encourage community or landowner partners to take leadership roles, larger engagement will be much more productive.
- You can either get things done or take credit for things getting done.
- Your partners will sell your program if you have invested time to build trust and credibility and helped them have "ownership" of the results.
- When possible, live within your focus area—being an actual member of the community tears down barriers and builds relationships.

- Seek out key leaders within your focus area—people who, if they adopt your program, will be ambassadors for it. Keep your ear to the ground, and the names will continue to come up.
- People are used to governments and bureaucracies not delivering, so do the opposite. Deliver in a timely manner to gain immediate respect for providing quality customer service.
- Engage yourself in the project and all efforts associated with it. Get your hands dirty!
- Stay in touch. If we have heard it once, we have heard it a thousand times: "Well, he [she] came out and talked about all he could do for me, and then I never heard from him again." Don't be that person.

When interacting with land stewards, minimize your verbal responses, as it encourages them to be open and forthcoming about expressing their goals, concern for their resources, land-use history, or explanation of an idea. In the conversation, be aware of your own nonverbal communications, which can be especially important. An affirmative head nod, a smile, direct eye contact, or body language that includes leaning toward the speaker are all cues that you are engaged and interested. Conversely, crossing your arms, looking away, looking down, or checking your phone are all nonverbal responses that suggest lack of attention or interest. This can be a tricky area, as there are cultures where direct eye contact or other forms of familiar nonverbal actions are not culturally acceptable. Hopefully, you will secure one or more mentors who will provide guidance for those situations.

Typically, we are not eager to share stories or experiences where projects, actions, or even ideas did not work out as planned or hoped, but these things are important to communicate as well. By keeping your failures to yourself, you may inadvertently allow others to experience those same mistakes. Be appropriately open and candid about your failures, especially if follow-up remediation actions are needed. Do not be embarrassed by failure, but instead learn from it and take advantage of the information in the next conservation effort. In the case of a successful venture, do not be bashful about sharing good examples of planned, designed, implemented, and/or monitored conservation actions. Theory, written pages, or academic examples are fine, but nothing beats seeing and sharing the management, restoration, or conservation practice applied to the land. There is no need for self-promotion, though. Instead, let your partners tell the story—they will be seen as sincere and believable among their peers.

Conclusion

As Ernest Hemingway wrote, "There are no stories inside four walls." His observation applies directly to our work in the "natural resources" world. It suggests that if we truly aspire to move away from being an academic or "book learned" conservationist, restorationist, or ecologist to one who is closely in touch with the land, then we must endeavor to spend as much time walking, understanding, and reading the land as part of a lifelong goal. It entails employing all five of our senses. What can be heard, seen, touched, smelled, or tasted? But that is just the beginning. What will you do with that information, knowledge, and perhaps later in your career, wisdom? How will you assess it? What actions will it lead you to when working with the producer? You will learn what Leopold observed:

> One of the penalties of an ecological education is that one lives alone in a world of wounds. Much of the damage inflicted on land is quite invisible to laymen. An ecologist must either harden his shell and make believe that the consequences of science are none of his business, or he must be the doctor who sees the marks of death in a community that believes itself well and does not want to be told otherwise. (Meine and Knight 2006, 265, quoting Aldo Leopold, *Round River*)

Leopold's suggestion is that each of us must become a Land Doctor. A thorough medical doctor will examine us in person using several diagnostic skills and/or order lab tests and procedures to assess all human systems (skeletal, digestive, muscular, nervous, circulatory, etc.) to ultimately assess our health. If your physician diagnoses health issues, he or she will try to prescribe a pathway (medication, diet, exercise, surgery, etc.) to regain your health. Like human beings, the land has numerous systems, processes, and cycles that are the same worldwide. Among them, water and nutrient cycles, energy flow, trophic levels (pyramids), plant succession and photosynthesis, biological diversity, and pollination come quickly to mind. As trained natural resource professionals, we can use our academic training (ecology, hydrology, ornithology, botany, pedology, etc.) and our practical experience (both empirical and observational science) to diagnose land health. Once problems are identified, we have an array of management, facilitating, and accelerating conservation practices available for use to prescribe ways for our land patient to restore health. It is at this intersection we also fully engage all our partners, governmental, academic, conservation districts, environmental and production groups, Extension, technical service providers, and most important, the land steward to plan, design, finance, implement, and monitor our work to see if we are moving our patient (the land) back to health. This is your charge as a natural resource professional, and having lived it, we hope these thoughts and checklists serve as a successful launch pad for your career.

CHAPTER 14

Land Stewardship

Professional Manager Perspective

BLAKE MURDEN, TY BARTOSKEWITZ, AND DECKY SPILLER

An emerging and growing sector of the natural resource community is the "professional land steward," who works for an individual or family who owns land. Given this intimate relationship, this chapter sheds light on the unique qualifications and skill sets needed to be a successful private-sector professional steward. Further, it discusses best practices for leveraging those skills into successful relationships that result in positive outcomes on the land.

Introduction

The professional land steward works in various industries across the United States in both the public and private sectors. This chapter focuses on private-sector professional land stewards and the skills necessary to succeed. Disciplines such as forestry, ranching, and conservation management may have position designations like general manager, natural resource specialist, wildlife biologist, ranch manager, or range specialist. Although these titles will undoubtedly vary by specific job descriptions, we collectively refer to them as professional land stewards for the purpose of this discussion. The objective of this chapter is to provide college students and early career professionals with information to help them understand the skills and attributes needed to become professional land stewards. Also, we hope landowners and hiring managers consider these skills and traits when developing job descriptions and interviewing candidates. We primarily discuss the types of land ownership one might encounter, thoughts on how to acquire a land steward position, and advice about how to succeed in the profession.

Land Ownership Types and Strategies for Conservation Success

There are a variety of ownership and business structures that employ professional land stewards:

- Individual ownership (one person owns and operates the property)
- Family ownership (multiple family members own and operate the property)
- Private operating foundation (a nonprofit entity, sometimes created by a family governed by a board of trustees)
- Private partnership (multiple owners or unrelated business entities sharing decisions)
- Corporate ownership (the property held as an asset by a company)
- Land trust (a trustee has legal ownership, stewardship, or partial control over a property at the behest of a beneficiary)
- Consulting role interacting with multiple owners or clients (working for oneself or a consulting firm, requiring interaction with multiple properties and owners)

Professional land stewards work with many different people, the most important of whom will always be the landowner or board of trustees who create vision, values, and job expectations and make the final decisions. Owners and trustees have varying personalities and backgrounds that will affect the level of finesse required to serve successfully in the land steward's role. Understanding the landowner's goals and objectives, dealing with differences in a professional manner, and providing best practice solutions are keys to success. Best qualities for working with any type of landowner include flexibility, clear communication, dependability, and honesty and integrity above all.

Flexibility is important in all areas of a professional land steward's career, whether it pertains to personal biases, work style, family and work activity schedule, or management techniques. Facilitating and implementing a landowner's vision may require one to set aside personal biases and instead rely on experience and knowledge to complete a task in any manner the landowner requests, as long as it is ethical and legal. There will be times when an owner will challenge the professional land steward's ideas. Although owners are generally open to hearing new perspectives, it is not the job of the land steward to win every debate. Instead, land stewards should always seek to find compromises that result in a win-win for everyone involved and move the project closer to the desired outcome. In all circumstances, land stewards should strive to provide educated, creative, and thoughtful solutions and ideas and the necessary steps to implement those into action.

Natural systems often require complex solutions, so both landowners and their professional land stewards must be adaptive when planning and applying management techniques. Together they must be willing to take calculated risks and use failures as learning experiences for future endeavors. Professional land stewards are tasked each day with making spot decisions without ownership guidance that affect operations. Some of the most common situations involve managing employees and guests, budgets, livestock and wildlife, and risk. Many land stewards operate under frameworks in which the day-to-day decisions, planning, and ranch operations are done without immediate consultation with an ownership entity. They operate, manage, and make decisions under an existing shared vision that provides a guide for conducting normal operations.

A unique challenge that natural resource managers and owners face is delayed gratification. Much of what they implement in their careers is influenced by natural processes and often will not come to fruition for many years. Implemented structural and natural resource management actions are remarkably similar to long-term business investments. Respectfully communicating this to landowners and maintaining a willingness for flexibility as situations change are valuable components to successfully realizing their vision for their land.

Effective and frequent communication between the landowner and professional land steward is necessary for successful project implementation and completion. Landowners may have vast experiential knowledge to share, so be alert to the opportunity to learn from them. Listen to the landowner's vision, goals, and objectives when beginning a new project, and *work together* to discern the specifics of each. If necessary, be prepared to humbly educate the landowner on topics as needed. Spending time in the field together is an excellent way to learn from each other and gives the landowner the opportunity to participate in the hands-on management activities. Inviting the landowner to take part in all phases of a project allows everyone to celebrate the positive results and collectively learn from those results that are less than desirable. Negative results create opportunities to share ideas, generate potential solutions, and foster a listening environment between employer and employee. In every situation, the professional land steward must clearly understand and represent the goals and vision of the landowner.

Integral to supporting a positive standing with their employer, land stewards must maintain their reputation in both the professional community and within the social community where they reside. Private land stewards are representatives of their employer, whether professionally or personally. Individual integrity and professionalism can have a marked impact on professional influence and success. Maintaining a reputation of integrity is a necessary practice for any professional, especially one who is part of the relatively small land management and natural resource community. The land steward should strive to be a person who presents solutions, resolves conflict, and continually seeks new sources of knowledge. Becoming a member of professional societies allows land stewards to continue professional development, learn and collaborate with their peers, and build a network of trust with fellow professionals. In addition to being members of the professional community, most land stewards work and live in small rural communities where personal character has a dramatic and long-lasting impact on the landowner's reputation, the land being managed, and opportunities for current and future employment. Therefore, striving to be a person of integrity and respect is paramount in this career field.

The New Professional Land Steward

Historically, most natural resource students came from rural settings like farms and ranches, which afforded them an opportunity to learn their job skills through experience. They not only learned practical skills, but their deep cultural background and close connection to the land likely framed their passion. Today, many candidates seeking land stewardship careers come from urban or suburban backgrounds. Experiences like hunting and fishing, high school agricultural classes, FFA and 4-H agricultural projects, or high school and college work experience and internships may have fostered a passion for the outdoors, land, and stewardship. These individuals have had a taste of what they see as their passion, but they will need to purposefully expand their perspectives and skill sets. Finding reliable, experienced mentors in their journey will be crucial to their success as a professional.

Typically, a bachelor of science degree in natural resources or agriculture sciences is necessary to qualify for this type of position, while some more advanced or specialized roles may require a master's or doctorate. The ideal professional land steward is a "jack of all trades" and should strive to become an expert in as many areas as possible. Furthermore, professional land stewards should respect and appreciate the rural lifestyle and its people, asking themselves before accepting a job if the local community is a good fit for them and their family. The professional land steward position is not a typical 8 a.m. to 5 p.m. job and will likely require working early mornings, late evenings, holidays, and weekends. Under such challenging conditions, one must hone skills for managing people and their expectations and be ready to make important, educated decisions with little time for preparation and planning. Given the complexities associated with people, livestock, and wildlife, the professional land steward will face interesting and rewarding challenges.

How to Get in the Gate

Landing a job as a professional land steward can take several years with many career-specific and seemingly unrelated job experiences that will over time contribute to a robust and favorable résumé. These experiences will help the novice land steward decide which positions and geographic areas are of interest (figure 14.1).

- Intern on private ranches, natural resource management companies, or state wildlife management areas across multiple ecoregions or environments.
- Volunteer to assist graduate students or professors on their research projects.
- Volunteer for any wildlife, range, and agricultural-related opportunities available through schools or communities.
- Get involved in professional and social organizations, and aim to hold leadership positions both at school and in a trade specialty.
- Find a mentor in the industry.
- Build a professional toolbox with a diversity of experiences and skills.
- Be willing to start at the bottom and view early career employment as growth opportunities.

Personal connections to landowners can lead to future job opportunities. A common path to full-time employment is to work for a landowner in a part-time, internship, or temporary role, thereby demonstrating one's abilities and work ethic. Once you "get in the gate," cultivating these relationships is essential, and, in most cases, this network will lead to full-time employment opportunities.

FIGURE 14.1. College intern assisting landowner with water system repairs (photo courtesy of Jonathan Vail).

Skills for Success

Professional land stewardship requires a certain set of basic knowledge and skills, and a qualified candidate must be able to demonstrate competence in both. Hard and soft skills are necessary to successfully face a variety of challenging topics. Hard skills such as economics, real estate, easements, mineral rights, risk management and liability, regulatory frameworks, property taxes, asset management, and writing proposals and grants are quantifiable and include the specific technical knowledge and abilities required. Soft skills are abstract, as they often relate to one's personality, emotional intelligence, or aptitude for interpersonal communication and cooperation, to name a few. These skills can and should be intentionally developed to enhance effective communication, listening, attention to detail, critical thinking, empathy, and conflict resolution. The following sections describe soft and hard skills that land stewards should consider and that natural resource employers are typically seeking.

Soft Skills

Interpersonal Communication

Communication in all forms is important for the professional land steward, including verbal, written, and body language and attentive and active listening. When conveying a message, land stewards should tailor the words to each specific audience. Land stewards will have to communicate effectively with employers, employees, other professionals, and the public. These situations require both managerial communication skills and the ability to use public speaking communication skills.

Collaboration and Negotiation

It is rare to find a position that does not require the land steward to collaborate and work well with others. Learn how to be a team player who can delegate tasks when appropriate, ask for help when needed, and build positive working relationships. An important aspect of collaboration is the ability to negotiate with others who have different goals or viewpoints. It is common for the land steward to be involved with negotiating on behalf of the landowner. Such negotiation may involve pipeline or transmission line right-of-way agreements, leases of various types (grazing, hunting, recreation, communication tower, wind turbines, solar array, etc.), and contractor or service provider terms and rates. You are a representative of the landowner and should negotiate in a thoughtful and respectful manner as the expectation and standard you set can impact the success of future collaborations and negotiations. The best negotiators always aim for everyone to walk away from a negotiation feeling like they gained something from the discussion and outcome.

Networking

Networking is one of the most valuable skills to learn as a professional land steward. When making introductions with other professionals, it is important to be sincere, make a good first impression, practice active listening, and ask relevant engaging questions. A successful networking interaction often results in the exchange of contact information and, more important, results in learning a few key facts about the other person that could benefit you in the future. Look for opportunities to practice networking skills by becoming involved in leadership positions in student organizations. Hone your communication skills by taking speech classes, participating on judging teams that require critical thinking and oral or written reasoning, or getting involved in student government. Interact with peers, professionals, and business leaders, and strive to emulate their best networking characteristics.

Critical Thinking and Problem Solving

The professional land steward must learn to approach problems logically and analytically. Finding solutions using a combination of both logic and imagination is a valuable skill. The best land stewards actively anticipate future problems and take steps to prevent them or mitigate their effects. They learn from past mistakes and offer multiple creative paths to achieve meaningful solutions to challenges. A calm demeanor and ability to turn a negative into a positive in every situation are extremely important.

Leadership and Management

The professional land steward must motivate staff, interns, volunteers, and other colleagues to work collectively and efficiently to achieve the goals and objectives of the organization. Creating expectations, assigning members to teams, delegating tasks, establishing deadlines, understanding and mitigating obstacles, monitoring performance and budgets, and setting good examples are all traits of an effective leader and manager. Working alongside employees in less-than-desirable job duties is a great way to be a leader and motivate a team.

Organization and Time Management

Managing your time efficiently and being organized are necessary for being productive. Properly prioritizing tasks aids land stewards in their ability to work efficiently, as balancing multiple unrelated disciplines in the span of a few hours is common for the professional land steward. Always have a plan to meet deadlines by deciding what is important, focusing on the project goals or milestones and endeavoring to complete them on time.

Professionalism

Maintain a professional image, and represent the owners and organization well, while integrating with the workplace culture of the organization. Develop the ability to interact comfortably with the landowner, family members, and guests, and be certain to uphold the goals and objectives of the owner both on and off the property. Lean on the credibility earned through education and experience when interacting with others, and maintain trust with the owner regarding confidential matters and situations. Emulate leaders who behave calmly and professionally, and take note of their behavior, dress, and associates. To use an old axiom, you "ride for the brand," and it is important to recognize that you are the organization's representative 24/7 on and off the property.

Risk Management

Recognize the weak links in processes and procedures, and mitigate potential liabilities for the landowner. Using networks of fellow professionals can help land stewards identify, monitor, and eliminate organizational blind spots. Although no plan is perfect and problems will still arise, these challenges create an opportunity for collaborative learning, communication, and plan modification.

Educational Acumen

A professional land steward should seek opportunities to invest in the future of land stewardship by educating youth, adults, and stakeholders. There will likely be occasions for the land being managed to serve as an outdoor classroom for students of all ages. School field trips, visits with local garden clubs, agriculture or natural resource field days, range-judging contests, and other similar events are all common opportunities for the land steward to teach about conservation, ecology, and agricultural resources (figure 14.2).

FIGURE 14.2. Blake Murden and Grant Lawrence inspecting a native grass restoration area dominated by yellow indiangrass (photo courtesy of Shield Ranch).

Hard Skills

Financial Knowledge

Professional land stewards have a fiduciary responsibility to the organizations and landowners who employ them to understand basic accounting principles. They should be competent in budget development and management and able to maintain and create balance sheets, profit-and-loss statements, complex project estimates, and business proposals. These skills allow land stewards to communicate effectively with the owner and financial professionals in their organization.

Legal Matters

Be familiar with state and federal wildlife and agricultural laws and cost-share programs that benefit range, wildlife habitat, and agricultural endeavors. Learn the laws related to conservation easements, mitigation banking, and especially real estate. Knowledge of real estate contracts and the process for land acquisition and disposition is essential. Additionally, understand basic legal terminology related to property rights, trespass, fencing, water rights, eminent domain, and the oil and gas industry.

Technological Advancements

Continually learn and adapt to new methods of communication. Although methods like email, video conferencing, PowerPoint presentations, and various social media platforms are common today, they are continually being updated and replaced. The same advancements are true for data analysis and recordkeeping programs like spreadsheets, statistics, graphics, and mapping software for GIS and GPS applications. Staying adept at current technologies will help you maintain your relevance and enhance communications. Be aware of current technologies like smartphone apps that can be used for field applications such as navigation, mapping, and plant identification, and share these helpful technologies with others.

Human Resources and Managing People

Many professional land stewards must hire and supervise staff, mentor interns, and manage contractors and service providers. However, few academic programs offer any formal education on this important organizational component. These people skills are invaluable when interacting with owners, stakeholders, industry peers, employees, and contractors. New managers may not immediately have the responsibility to oversee others and their hiring, but a good set of interpersonal skills needs to be honed throughout one's career. Effectively managing others begins with clearly articulating the employee's job responsibilities to establish a performance baseline. Finding and hiring the right people who demonstrate aptitude for the necessary skills and then playing to their strengths and offering support for their weaknesses will set the staff up for success. Find ways to make your employers and employees successful.

Natural Resource Management

The professional land steward should have an in-depth knowledge of ecology, the natural history of plants and animals, habitat management, and field skills that can be applied across many habitat types. For example, modern agricultural practices can be employed to enhance the environment and achieve the landowner goals. If knowledge of livestock management and farming practices is lacking, seek opportunities to gain these skills, as they will benefit the manager's holistic understanding of land stewardship.

Business Acumen

It is important to know how to act in a professional business meeting. Best practices are to take good notes, ask educated and thoughtful questions, and be a student of the greater business community. Land stewards should educate themselves on the business practices of ownership and understand their role in the organization's structure. Be professional and be humble, know your limits, and be cognizant of the expertise around you. You must be able to recognize when you can contribute and when you need to seek the advice and expertise of others within your network.

Entrepreneurship

Although often absent from conservation or agricultural science classroom teachings, entrepreneurship is a valuable skill for generating income using a property's existing resources, including wildlife, forage, infrastructure, or recreational opportunities. Both consumptive uses, such as hunting, fishing, grazing, or forestry, and nonconsumptive uses, such as ecotourism, can appeal to the public and generate additional income for the landowner or organization.

Advice for Students and Early Career Professionals

Individuals completing their education or beginning their careers often must learn from their own mistakes and unfortunate outcomes; however, there are extensive networks of natural resource professionals and land stewards who can share their own life experiences to mentor the next generation. Some valuable lessons that will help new professionals build a steady career more quickly might include the following:

- Build a network of trusted friends, colleagues, and experts.
- Join and stay active in professional societies and organizations.
- Remain respectful and open to continual learning.
- Seek out internships; they are a wonderful way to learn new skills and interact with new people and will often lead to a full-time position.
- Embrace the idea that the experiences and skills gained from an entry-level position will provide the expertise needed for future advancement.
- Broaden your horizons; be open to gaining new experiences and perspectives by working in other states, regions, or countries.
- Ask questions often and be able to summarize the answers returned.
- Listen and stay active throughout the process of project planning and execution.
- Be memorable for the right reasons; be polite and respectful, have a positive attitude, and display excellent work ethics.

Conclusion

Many private-sector land stewards live and raise their families on some of the most amazing and scenic properties in the nation. Often they are provided housing, utilities, office space, vehicles, and access to outdoor recreational amenities. Most have some combination of retirement planning, health insurance, profit sharing, incentive bonuses, recreational opportunities, or other benefit programs that add to the employment package. Being a professional land steward affords the opportunity to make one's vocation and avocation one and the same. The diverse job responsibilities, attractive compensation packages, and the opportunity to make long-term positive changes to the land, and the people who enjoy and benefit from well-managed natural resources, make the land steward role an extremely rewarding career choice. We hope the experiences and advice we have shared in this chapter will help students and early career professionals realize a fulfilling role as a successful private-sector land steward.

CHAPTER 15

Land Stewardship

Landowner Perspective

LANDOWNER TESTIMONIALS

> A book about private land stewardship would be incomplete without a chapter written by private landowners. While stories and quotes from many of these landowners have been sprinkled throughout the book, this chapter gives them a specific platform to share their experiences and perspectives. Astute natural resource professionals will recognize valuable advice herein and apply it to their daily work.

Introduction

It is no secret that the job of a natural resource professional is a complicated one, especially in the private land states that are the focus of this book. On the one hand, you are charged with meeting specific goals and promoting practices important to the mission of your agency or organization and, on the other, with servicing the needs of landowners with diverse objectives, challenges, and operational characteristics. To create a positive benefit for both, what is the recipe for working effectively with landowners?

In the previous two chapters you learned about the importance of collaborative relationships with land stewards and how to foster them from the perspective of seasoned natural resource professionals. From their guidance and experiences, we hope that you will avoid some common pitfalls and learn from their many successes to master the skill of developing relationships that lead to conservation outcomes on private land.

While the previous chapters probably could have stood alone and will certainly serve as a guidebook for many, we wanted to take this exercise one step further—through the eyes of the landowner. As you learned from the practitioners in the preceding pages, the best way to understand landowners, both individually and collectively, is by hearing directly from them. To that end, we partnered with the Sand County Foundation and interviewed Leopold Conservation Award recipients from each of the regions identified

FIGURE 15.1A. Through our partnership with the Sand County Foundation, we interviewed landowners from each of the regions identified in chapter 6, each representing a wide range of demographic and operational characteristics.

The Leopold Conservation Award Program recognizes agricultural landowners actively committed to a land ethic. Working with prominent state conservation partners, Sand County Foundation presents the prestigious honor, which consists of $10,000 and a crystal award, in settings that showcase the landowner's achievements among his or her peers.

The Leopold Conservation Award is a productive investment in private land conservation. It recognizes and celebrates extraordinary achievement in voluntary conservation by private landowners, inspires countless other landowners by example, and provides a prominent platform by which agricultural community leaders are recognized as conservation ambassadors to citizens outside of agriculture. Finally, the program builds bridges among agriculture, government, environmental organizations, industry, and academia to advance the cause of private land conservation.

in chapter 6 (see figure 15.1a), attempted to synthesize their thoughts here, and share their words as a source of inspiration, insight, and validation to drive home the concepts from the previous chapters. In the subsequent pages, we share the words of these land stewards with you directly so you can gain an appreciation and understanding of how to best engage with private landowners.

Landowner Interviews

Due to the nature of stewardship and propensity toward collaborative conservation that is characteristic of Leopold Award recipients, these interviews yielded a treasure trove of perspectives that we hope will serve to inform many productive relationships into the future.

Our hope is that by sharing the thoughts of these outstanding and thoughtful land stewards, in addition to the practitioners from the previous chapters, private land conservation will be elevated and accelerated through the work of future land stewards, managers, and their agency counterparts.

BRE OWENS | COBBLESTONE RANCH | CHICO, CALIFORNIA

Bre Owens grew up in a ranching family and now owns her own piece of California, which is primarily a cow-calf operation. She also raises Dorper sheep on 175 acres of rangeland she owns and several grazing leases she maintains on Federal Refuge lands. Bre's stewardship philosophy is one of using livestock and active management to promote desired vegetation response and improve ecosystem health. When asked about best practices for promoting a strong working relationship between landowners and agency practitioners, Bre listed the following key elements:

FIGURE 15.2. Bre Owens (photo courtesy of Bre Owens).

1. **Listen.**

   The biggest thing is to listen, and don't expect to have all the answers. So often, we put these young people in the role of a technical assistance provider or technical service provider as if they're supposed to totally understand the system and situation each landowner is in. Then we expect them to have all these recommendations. It's an unfair expectation for the young person, and it's really not even what most landowners want or need.

2. **Ask questions.**

   Approach conversations with landowners as an exploration or brainstorming session. By allowing the landowner to think out loud and think through some of these things and explore some things on their own just by having that conversation. This approach is much more effective than some young person coming in with a bunch of ideas or solutions before they really understand the situation the landowner is in.

3. **Match the landowner's work ethic.**

   Land ownership is not an 8–5 job—you may have to meet with them at odd times to match or accommodate their schedule. Young practitioners also need to understand the complexity of working the land and that unanticipated challenges and obstacles pop up on a daily basis that impact how land management is done.

BRIAN TREADWELL | TREADWELL CATTLE CO. | FT. MCKAVETT, TEXAS

Brian Treadwell of Treadwell Cattle Co. (TCC) operates a free-range commercial hunting business and cow-calf operation at his 8,000-acre historic family ranch, settled in 1887 near Ft. McKavett, Texas. The ranch is situated at an elevation of 2,200 feet on the Edwards Plateau, where rolling flat-topped hills covered in live oaks have been carved by centuries of flash flooding, and averages 22 inches of rainfall each year. TCC also manages two irrigated stock farms on the South Concho River near Christoval, where the Treadwells maintain two commercial Dorper sheep flocks and grass-finished beef from the ranch using Bermuda grass interseeded with legumes. Treadwell's objective is to improve the landscape through stewardship and views the operation holistically.

FIGURE 15.3. Brian Treadwell (photo courtesy of Brian Foster).

When we asked Brian what kind of advice he had for young agency personnel hoping to work effectively with private landowners, he said this was his best advice:

1. **Listen and observe.**

   Use your ears and eyes. Set your ego and all the cool stuff you've done to date aside. The mentor in your presence has likely forgotten more cool stuff than college provided you. Only talk when seeking clarity. Always ask questions about how it used to be—looking back at the history of the rangeland should help define its future potential.

2. **Show, don't tell.**

   Fielding Harwell was our Texas Parks and Wildlife Department biologist for many years, and he was the one who helped me understand the value of fire to restore our landscape. But I didn't really understand until he took me to SEE the results of management with the use of prescribed fire at one of the local wildlife management areas [WMA]. The properties on both sides of the road leading up to the WMA looked just like our land—sparse grass and a lot of rocks and juniper, but the moment we drove through the gate of the management area that had been managed by fire for many years, it was a sea of diversity like I had never seen before. We started burning later that year and we've never stopped!

3. **Let your passion shine through.**

   Every effective agency person I have worked with had a real passion for their job. I always got the sense that it didn't matter what they were getting paid; they really cared about the work they were doing. If you have that kind of passion for natural resources management, it will be contagious, and landowners will want to work with you.

DICK CATES | CATES FAMILY FARM | SPRING GREEN, WISCONSIN

Dick Cates has a PhD in soil science and spent the years of his early career working on saline soils in Saudi Arabia, but he never got "farming out of his blood," so he came back to run the family farm as soon as he could. Dick and his wife, Kim, now run a grass-fed beef operation on their 900 acres in the Driftless area of Wisconsin. Their farm is also home to a cold-water trout stream that they have restored over the years through managed grazing and partnerships with agencies. The stream is now a Class 1 trout stream, and the Cates family opens their gates for field days and research to share what they have learned through years of restoration work.

Through Dick's extensive experience working with a number of agencies, he brings a unique perspective to effectively work with landowners:

1. **Listen and ask questions.**

   I feel like the best agency representatives are those who provoke dialogue amongst landowners.

2. **Show, don't tell.**

   Understand that people's motivations are all different, and you're never going to get anywhere by telling them what it is they should be doing. The best ones I know help us understand what we could be doing by showing us examples. Putting on field days that show good examples is so much more powerful than just telling us what to do.

3. **Work to build a sense of community among landowners.**

   The community piece is huge. When the USDA folks started a watershed group many years back, most people came to the meetings with the attitude of "you're not going to tell me what to do on my farm!" But over time, by giving them the forum to do so, the landowners began to share notes and learn from one another. Now they all try to outdo each other with their successful management stories! And none of it would have happened if they hadn't been brought together by an agency person.

FIGURE 15.4. Dick Cates and Jersey steer (photo courtesy of Sand County Foundation).

DEAN JACKSON | MT-GLEN FARMS | COLUMBIA CROSS ROADS, PENNSYLVANIA

Dean and Rebecca Jackson, along with their four children, own and operate the 925-acre Mt-Glen Farms, a registered dairy cattle operation in the Chesapeake Bay watershed in northeastern Pennsylvania. Their vertically integrated operation includes crops for feed, homegrown and nationally marketed Holstein genetics, and milk sales. They also participate in the Conservation Reserve Enhancement Program through the USDA Natural Resources Conservation Service to set aside sensitive riparian areas to benefit watershed health. The Jacksons have enrolled 30 acres into a 15-year contract with USDA, illustrating their passion for conservation and stewardship. Dean describes his stewardship ethic this way: *"It's being a caretaker with a deep joy and love and passion for making that land the very best it can be so that I can pass it on to the next generation in better shape than when I got it."*

When speaking of the important relationships he has had with agency personnel during his land stewardship tenure, he describes these important factors that led to positive outcomes:

1. **Use communication that leads to a relationship.**

   Nathan Dewing is the head of our Bradford County Conservation District, and I believe he has had so much success in restoring this watershed because he takes the time to communicate with the people and he has really built strong, lasting, trusting relationships with the people of this county. He isn't seen as an outsider; he is a part of our community, so everyone trusts him.

2. **Be honest and upfront.**

   If a person comes on this farm, I want them to be upfront with me. Some of these kids try to act like they know more than they do, and that is always a turnoff. On the other hand, if they are completely honest and say, "I didn't grow up on a farm and I have a lot to learn, but I am passionate about doing a good job and serving you however I can," I'm going to really respect that.

3. **Show some grit.**

   Farmers have a lot of grit, and we like to see grit. If a farmer rejects you and tells you he's not interested in working with you, show some grit. Let a month or two pass and try it again. Don't give up on that relationship.

FIGURE 15.5. Dean Jackson (photo courtesy of Sand County Foundation).

JAY TANNER | DELLA RANCHES | GROUSE CREEK, UTAH

Jay Tanner and his family run cattle on almost 800,000 acres of owned and private lease and BLM land in the Great Basin area of northwestern Utah. The Tanner family has been ranching in this area for five generations, and Jay describes his stewardship philosophy this way: *"As a steward, I intend to make a living off of the land today, but I also keep an eye to the future for myself and for those people who will hope to utilize these properties in the future to make a living. I also realize that what I do with my land has an impact on the other people around me, my neighbors, and the climate."*

While Jay is mostly focused on livestock production as a source of his livelihood, he sees wildlife as an indicator of range health and therefore considers that wildlife conservation can be a win-win for production agriculture. He says, *"The wonderful thing is that the sage-grouse are using the areas where we're removing the juniper, and it's made the areas significantly better for other species of wildlife and livestock as well."*

FIGURE 15.6. Jay Tanner (photo courtesy of Stuart Ruckman Photography).

Jay has worked extensively with agencies, wildlife conservation groups, and researchers helping to develop partnerships and promote conservation of sage-grouse, as well as a species of trout, the Lahontan cutthroat, previously thought to be extinct until it was found in a stream on his ranch. When asked about the key elements of a productive relationship between an agency representative and a landowner, he listed the following:

1. **Listen and observe.**

   People skills are important. If they will come in and take a little time at the kitchen table with ranchers and learn what their situation is, that builds trust that's really important. When you develop that relationship first, then you can be more effective when it's time to say, "What if we try this, or how could we achieve this goal together?"

2. **Become a part of the community—understand the culture.**

   Some of the most outstanding public land managers I have dealt with have taken the time to sit down and discuss issues and really get to know the area and the culture before they start trying to change things or make recommendations.

3. **Gain experience.**

   I have found that many of the recent college graduates get out of school and get jobs in the field working with landowners before they have ever had any real-world experience on a ranch or farm. My advice would be to seek out opportunities to gain that experience while you're in college, or even after, so you can have a better understanding of the day-to-day operational challenges of ranching before you start working with landowners. I encourage other landowners to open their gates and doors to interns or college researchers to gain those experiences as well. You never know where they'll end up. Some of the kids that have interned or done research on my ranch are now high up in natural resource agencies, and that is a good feeling as a landowner—to know you might have contributed to the professional development of someone who is now making a great contribution to agriculture, ranching, and conservation.

JIM HEBBE | HEBBE FARM | GREEN LAKE, WISCONSIN

Jim Hebbe and his wife, Valerie, were pioneers of no-till farming in east-central Wisconsin, where they operate a 1,300-acre grain farm. As an early adopter of conservation tillage, Jim acknowledges his role as an ambassador of these more ecologically responsible farming practices and has been happy to set an example to other farmers. Jim said, *"There is a lot of peer pressure in farming. Your work is always on display for everyone to see, so I am always looking to other farmers for what they have learned works and what doesn't. When we started having success with conservation tillage, both ecologically and economically, others started to take notice and followed suit."*

In addition to being a full-time farmer, Jim, who has an agriculture economics degree with a minor in agronomy, also served as the land conservation director for Green Lake County for 30 years. Having been on both sides of the practitioner-landowner relationship, Jim offers a unique perspective on best practices for collaborative conservation with landowners:

1. **Have a conversation before undertaking conservation.**

   You need to talk to that landowner and see where they're coming from. Identify your areas of common philosophy and beliefs and work from there.

2. **Remember who pays your check.**

   I always worked really hard to communicate to landowners I worked with that it's not just farmers who pay my wage. It's also the general public downstream who want water quality, wholesome food at a reasonable price, etc. On the other end of the spectrum, I would say to nonfarming people I interacted with that the reality of farming was much different than they sometimes assume, and that livelihoods were at stake, so we have to be careful and deliberate about what we ask of them.

3. **Be honest and transparent.**

   Again, this goes both ways—on the landowner side and the practitioner side. In order to have a good working relationship, you have to be genuine, honest, and open about what concerns or issues that you have or that are important to you, and then be open to hearing and accepting other points of view.

FIGURE 15.7. Jim and Valerie Hebbe (photo courtesy of Sand County Foundation).

SHARON AND PAT O'TOOLE | LADDER RANCH | SAVERY, WYOMING

Sharon Salisbury O'Toole's great-grandparents established the Ladder Ranch in the Little Snake River Valley along the Wyoming-Colorado border in 1881, where six generations have lived and worked. The family, guided by a strong generational stewardship ethic, uses rotational and bi-species grazing of sheep and cattle. The family seeks to balance production and conservation and are always focused on long-term sustainability. Sharon's husband, Pat, served as a Wyoming state representative and board member for Partners for Conservation (now Partnerscapes). Pat understands the value of collaboration as well as any landowner. The Ladder Ranch is home to several "species of interest." Pat credits the variety of agencies and organizations they work with for the family's ability to balance their need for production with their desire for conservation of these species.

FIGURE 15.8A. The O'Toole family at the Ag Hall of Fame, 2022 (photo courtesy of Sharon O'Toole).

FIGURE 15.8B. Pat O'Toole (photo courtesy of Sharon O'Toole).

When asked about their thoughts on best practices for an effective landowner-agency relationship, Pat and Sharon listed the following key elements:

1. **See landowners as the solution.**

   Most landowners try to do what's best for the land—we shouldn't have to fight the government to do it. . . . We had one of the best BLM guys for 40 years. He was as good a guy as God makes. I always knew I could trust him because he wanted what was best for the land too, and saw us and other landowners as the solution.

2. **Understand the stakes.**

   Agency folks need to realize that if they make mistakes, they still get a paycheck. Sometimes landowners are a little more cautious about adopting new practices or jumping into programs because if it doesn't work, it could affect our livelihood. The best agency practitioners I've known have understood that and have been very patient with the landowners they work with.

3. **Make it your mission to do good.**

   There is no better career where you can feel like you really accomplished something that will make a difference in the world than helping landowners steward our natural resources. We need people who want to do good, who want to succeed and want to build lifelong relationships based on conservation.

Prior to the publication of this book, the O'Toole family and the Wyoming ranching and conservation community lost a patriarch, a champion, and a thought leader with the untimely passing of Pat O'Toole in February 2024. How appropriate that he ended his interview with us with the quote above. May we all wish for a legacy as strong as that of Mr. Pat O'Toole.

STACY HOEME | HOEME FAMILY FARM & RANCH | SCOTT CITY, KANSAS

Stacy Hoeme and his son, Chaston, operate approximately 20,000 acres of cropland and shortgrass prairie rangelands in western Kansas. Testament to the Hoeme family's stewardship is the thriving dual population of both greater and lesser prairie-chickens that inhabit their land in concert with their production agriculture business. In fact, the discovery of these populations on the Hoemes' land, and the family's effective interaction with multiple agencies and organizations, in addition to the habitat on their ranches, are what ultimately prevented listing of the species.

The Hoemes have hosted numerous wildlife research projects that have both improved their understanding of their impacts on habitat and contributed to the greater body of literature on many subjects. Based on Stacy's extensive partnership efforts with many conservation agencies and organizations, he gives the following thoughts on how to foster effective landowner-practitioner relationships that will yield mutually beneficial results:

1. **Be truthful and honest about your knowledge level.**

   If you don't understand something, don't try to blow smoke about it; just say, "Hey—tell me more about what you're doing," and I'll be happy to help you understand.

2. **Be willing to spend time to build a relationship.**

   You can tell if someone's not quite into the conversation—kind of listening with one ear—and not sincere about their interest in what you're doing. I like it when folks show me some personality so I can get to know them. We live way out in the middle of nowhere; it's nice to have people come out that end up being friends.

3. **Listen.**

   We may not have fancy college degrees, but we're out here every day observing what's going on, so sometimes we are able to bring perspective about things they haven't thought of. We had some researchers studying deer out here, and they were getting weird results. I finally told them that they were disturbing those deer by tracking them every day and they probably weren't acting normally because of that. They hadn't even thought of that.

FIGURE 15.9. Stacy Hoeme (*left*) and son, Chaston Hoeme (*right*) (photo courtesy of Sand County Foundation).

KIMBERLY RATCLIFF | CANEY CREEK RANCH | FAIRFIELD, TEXAS

FIGURE 15.10. Kimberly Ratcliff (photo courtesy of Kimberly Ratcliff, Wynn Myers Photography).

Kimberly Ratcliff left a successful career on Wall Street to come back home to Texas to manage her family's 2,500-acre diversified ranching operation in east-central Texas. The family runs a certified Carbray herd, manages their own feedlot, and has recently launched "Farm to Freezer," a business that direct-markets their beef to the public. They additionally make their own silage, harvest their own hay, and plant grass sprigs for other area landowners. In addition to this highly successful vertically integrated agricultural operation, Kimberly started a community program called "100 Ranchers," where area landowners get together and share ideas and cooperate on projects that are difficult for one small landowner to do on his or her own.

One of the great successes of 100 Ranchers has been connecting small, minority landowners with agency personnel and service providers. Kimberly is passionate about helping other landowners build strong business models into their operation and creating a platform for collaboration and interaction with agencies. Her thoughts on successful landowner-agency interactions follow:

1. **Listen.**

   Listen more than you talk. Be a resource. Producers many times don't know where to go for information or resources, so be willing to guide them to the right place even if you don't know the answers or the problem doesn't fall within your job.

2. **Get experience.**

   I always say it falls on ranchers and producers to prepare the next generation for what college might not be able to provide them. I go through Texas A&M Extension and try to provide internship opportunities for kids—especially minority women like me—while they're in college. I find that it is so important to get that experience BEFORE you graduate. If you're just studying for agriculture, and then you get hired by NRCS and have never actually worked on a farm or ranch, you are not going to be effective at working with producers.

3. **Help build community.**

   I have never had a problem relating with agency folks, but one of the reasons I started 100 Ranchers was that I was seeing that many of my fellow minority landowners were a bit distrustful of the folks at the natural resource agencies. Once we built 100 Ranchers, it kind of created a sense of community that allowed us to interact with the agencies as a group, which made everyone more confident. Now, when NRCS, as an example, wants to work with someone in our group, they come to me first, and I, as someone the producer feels more comfortable with, am able to make the introduction and help to build that trusting relationship.

REESE THOMPSON | VIDALIA, GEORGIA

Reese Thompson is a sixth-generation Georgia landowner whose family property is located in the flat to gently rolling hills about three hours south of Atlanta. Reese says, *"We plant and manage longleaf pine trees for pulpwood, saw logs, and poles. I manage several thousand acres for my family. Growing trees is both an art and a science. It is an art to enhance the beauty of the forest and improve wildlife habitat. It is a science to produce the most tons of fiber per acre of the most valuable product."*

Reese serves on the boards of several conservation organizations and works closely with numerous groups to promote education about prescribed fire and conservation of sensitive species like the gopher tortoise. He is the creator of Burner Bob, a bobwhite quail mascot that promotes the wise use of prescribed fire to create good habitat for sensitive species. Burner is a counterweight to Smokey Bear.

Reese's strong working relationships with so many groups and organizations has led him to provide the following simple advice to young natural resource practitioners. It sums up well what we have heard from so many of these land stewards:

> To a new biologist, my best advice would be to visit as many landowners as possible. Spend twice as much time listening as talking, with an emphasis on asking questions. Given time, the landowner will express his objectives and problems.

FIGURE 15.11. Reese Thompson (photo courtesy of Reese Thompson).

Conclusion: The Recipe

For natural resource professionals, it is sometimes easy to operate in a way that places internal reporting metrics ahead of mission and outcomes. Here we learned directly from the landowners we interviewed that it is critically important to meet them where they are and operate within their established relationships and/or comfort zone(s) rather than try to force something new on them from an angle that gives them pause. This is where listening and relationship building are key to developing the foundation for future successes.

Dean Jackson, the Wisconsin dairy farmer interviewed, said, "*When someone new comes along, it's just kind of natural for red flags to go up. Your guard is up when you see a government truck pulling in the driveway. Takes time to break down those barriers. It's okay to acknowledge that they might be starting with some preconceived notions about you.*" Building trust is a paramount first step that cannot be overlooked.

Preconceived notions may be about the agency rather than the person, presenting another obstacle to overcome. For example, landowners may have a strong wildlife focus on their property and may participate in state-level incentive programs, yet they may be wary of federal endangered species programs that they perceive as not being in their best interest. In this case, partnerships can be developed to break down these barriers by linking state personnel with whom they are already comfortable and thereby building trust in the program over time.

This recommendation about collaboration speaks to this reality that sometimes agency personnel must be willing to take a back seat and do what it takes to get the greater good accomplished, even if you or your agency doesn't get the "credit." If you need a landowner or group of landowners to cooperate with you or your programs, but they seem wary, learn about them. Find out whom they trust. If they are more likely to talk to and work with another agency or private service provider, work with that entity or person to develop programs or partnerships that are more palatable to the stewards you need on board.

When we set out to interview these landowners, we anticipated that there would be some overlap in what they told us, but we had no idea how similar the advice and needs would be from such a diverse group of people. The recipe is really simple. Listen more than you talk, be sincere, be real, respect cultural and personal norms, focus on relationships first, be adaptable, be trustworthy and accountable, nurture partnerships, and develop a sense of community. In the end, landowners want their agency representatives to be real people who can share their desire to leave the land better than they found it and provide them support and tools that will help them achieve their goals.

CHAPTER 16

Spiritual Responsibility of Stewardship

GENE T. MILLER, MATT WAGNER, AND OSWALD JOHN NIRA

With case study provided by Craig D. Galli.

Many land stewards view care and management of the land as much more than a vocation or way of life; instead, it is truly a spiritual calling. In fact, many religious perspectives include a spiritual call to stewardship within their doctrine. This chapter discusses this calling through the lens of global religion and offers perspective on the practice of stewardship as a vehicle for individuals and groups to demonstrate their spiritual calling on behalf of the Creator and for the benefit of future generations.

Introduction

The concept of stewardship has been widely discussed in recent years, and here we define it according to *Merriam-Webster* (2021) as the "careful and responsible management" of goods entrusted to one's care. Contemporary discussions contextualize *stewardship* within a historical and cultural milieu; thus, an understanding of the concepts of *goods* and *entrust* become necessary for a holistic view and understanding of the term *stewardship*. The spiritual responsibility of stewardship necessitates, for many, the awareness of the manner through which the structures of stewardship were developed and how this manner shapes care for the earth and its inhabitants. Stewardship may be viewed from multiple perspectives, including, for example, colonialism and/or abuse or use of resources for personal gain and profit. Other perspectives consider stewardship within a framework of specific relationships, cultures, and spiritualities, in which actions in relation to the land are motivated by one's understanding of a responsibility and "calling" for an appropriate action or response. This chapter introduces the reader to these latter perspectives: a survey of various spiritualities and their relation to the land and its resources. However, it is important to note the former set of perspectives, which may shape and influence the latter in certain cultures and spiritualities. Here we present a framework for why landowners or managers may be strongly motivated from a spiritual standpoint to pursue stewardship of land, water, plants, animals, landscapes, and ecosystems, irrespective of time and financial requirements, and in many cases for the pure satisfaction of "leaving it better than they found it."

Christianity

Many land stewards are guided and motivated by a strong belief in God and Creation, where God created what we call Nature and gave humanity the responsibility of caring for this Creation. Individuals take their role as stewards and caretakers of the land very seriously and consider it a God-given responsibility. Many land stewards consider themselves to be temporary custodians rather than landowners. The Bible states, "The land must not be sold permanently, because the land is mine and you reside in my land as foreigners and strangers. Throughout the land that you hold as a possession, you must provide for the redemption of the land" (Leviticus 25:23–24). Those who advise and assist landowners should learn to appreciate this perspective and not underestimate the role of faith in inspiring a land ethic. Aldo Leopold, widely considered to be the father of wildlife management in the United States, studied the Bible for clues to the origin of forestry and our relationship with the environment. He wrote, "There are two things that interest me: the relation of people to each other, and the relation of people to the land" (Leopold 1947, 337).

Stewardship of Creation

Many Christians have the belief that "God is in control, not us," and that in response to Creation's majesty, we are inspired to be stewards of it. It is common to see land stewards humbly express they are a part of Creation but not the master of it. The Bible explicitly points this out when Job was questioned: "Where were you when I laid the earth's foundation? Have you comprehended the vast expanses of earth? Do you know when the mountain goats give birth? Do you watch when the doe bears her fawn?" (Job 38–39). These questions from God humbled Job and put things in perspective regarding his place in the universe.

It is important to note that many natural resource professionals who assist landowners may also have these same beliefs and motivations. An encounter with a rancher illustrates this point when he stated, "As a farmer and rancher, I tend to believe that if I will work harder and work longer, I can make the land produce what it should. In truth, I cannot make the rain come or the sun shine. I have to remember that that comes from the Lord in his own timing." Land stewards may be more apt to recognize a dependence on a higher power in their daily activities, including the mystery of birth, the reality of mortality that, for example, may stem from witnessing disease or predation, the vivid splendor of sunrises and sunsets, the intricate weaving of the soil and the flora and fauna, or the feeling of smallness when observing the constellations.

Applying a Stewardship Ethic

Several passages in the Bible illustrate how a stewardship ethic develops. For example, fire has been one of the most important ecological processes in the formation of natural landscapes. Historically, wildfire shaped the land, removing old growth and releasing fresh grasses and flowering plants. Native Americans used fire to rejuvenate vast grasslands, attracting bison and other wildlife that could then be hunted for sustenance. Land stewards today understand that prescribed fire can facilitate restoration of native grasslands lost to brush infestation. Here is one biblical example that instructs the application of this stewardship practice as a spiritually driven process, where fire is used to rejuvenate and "cleanse."

> The ax is already at the root of the trees, and every tree that does not produce good fruit will be cut down and thrown into the fire. I baptize you with water for repentance, but he who is coming after me is mightier than I. He will baptize you with the Holy Spirit and fire. (Matthew 3:10–11)

Connection to the Natural World

Throughout the Bible, there are references to the relationship of humanity with the natural environment, from the importance of watershed management (Joel 3:18), to weed control (Matthew 13:24–30), development in rural areas and provision of open space (Isaiah 5:8; Ezekiel 45:2), forestry (Isaiah 41:18–19, 44:14; Ezekiel 17:22–24; 31:3–9), and other conservation topics. Following are some examples of the connection between humans and their natural environment:

> Then God said, "Let us make mankind in our image, after our likeness, so that they may rule over the fish in the sea and the birds in the sky, over the livestock and all the wild animals, and over all the creatures that move along the ground." (Genesis 1:26)

> The Lord God took the man and put him in the Garden of Eden to work it and take care of it. (Genesis 2:15)

> Everything that lives and moves about will be food for you. Just as I gave you the green plants, I now give you everything. (Genesis 9:3)

These biblical references establish the premise that God as Creator established the earth and all things in it. It goes on to describe the various forms and activities of stewardship of land and its natural resources and how we are connected to this natural environment. We understand that human beings are part of a system, part of our planet. The Bible also describes other benefits of the natural world beyond consumptive uses. Benefits such as solitude, refuge from the chaos and constant noise in our lives, can be obtained through "quiet places," allowing one to decompress and meditate. Here are some examples:

> Now when Jesus heard this, he withdrew from there in a boat to a desolate place by himself. (Matthew 14:13)

> And he said to them, "Come with me by yourselves to a quiet place and get some rest." (Mark 6:31)

> But Jesus often withdrew to lonely places [wilderness] and prayed. (Luke 5:16)

> At once the Spirit sent him out into the wilderness, and he was in the wilderness forty days, being tempted by Satan. He was with the wild animals, and angels attended him. (Mark 1:12–13)

This connection to the natural world instills a sense of marvel for many people. Similarly, land stewards enjoy the solitude of their lands and delight in sharing those lands with family and friends, whether from a duck blind, behind bird dogs on point, at the base of a large tree while waiting for a wild turkey to appear, bird-watching, or enjoying the glorious panorama of a sunset from a front porch in "stillness."

Stewardship as Process

The book of Leviticus is instructive for land stewardship by introducing the foundational concept of land restoration that is so critical to maintenance of natural landscapes for present and future uses and is worth repeating:

> The land must not be sold permanently, because the land is mine and you reside in my land as foreigners and strangers. Throughout the

> land that you hold as possession, you must provide for the redemption of the land." (Leviticus 25:23–24)

This passage conveys that land stewards are not permanent landowners in God's time but instead temporary caretakers of land, water, and other natural resources for the benefit of all humanity. This reference charges the reader with "redemption of the land," which means to restore the land, and explains the origin of the term *renewable natural resources* that is used today. The following passages also establish the concept that with neglect and careless management, natural resources will suffer and, in time, humans that depend on them. This leads to the reasonable conclusion that careful stewardship of natural land, waters, and wildlife is not optional or permissive but necessary and worthy of humanity's efforts. *Redemption* or restoration is not only needed, but God has those who have temporary care of the land make sure that it is accomplished.

> I brought you into a fertile land to eat its fruit and rich produce. But you came and defiled my land and made my inheritance detestable. (Jeremiah 2:7)

> The earth dries up and withers, the world languishes and withers, the heavens languish with the earth. The earth is defiled by its people; they have disobeyed the laws, violated the statutes and broken the everlasting covenant. Therefore a curse consumes the earth: its people must bear their guilt. Therefore, earth's inhabitants are burned up, and very few are left. (Isaiah 24:4–6)

The final reference illustrates the process of stewardship through a myriad of feelings, traits, and practices that together strengthen a steward's motivation, drive, and confidence to do "the right thing":

> Praise the Lord, my soul. Lord my God, you are very great; you are clothed with splendor and majesty. The Lord wraps himself in light as with a garment; he stretches out the heavens like a tent and lays the beams of his upper chambers on their waters. He makes the clouds his chariot and rides on the wings of the wind. He makes winds his messengers, flames of fire his servants. He set the earth on its foundations; it can never be moved. You covered it with watery depths as with a garment; the waters stood above the mountains. But at your rebuke the waters flowed, at the sound of your thunder they took to flight; they flowed over the mountains, they went down into the valleys, to the place you assigned for them. You set a boundary

> they cannot cross; never again will they cover the earth. He makes springs pour water into the ravines; it flows between the mountains. They give water to all the beasts of the field; the wild donkeys quench their thirst. The birds of the sky nest by the waters; they sing among the branches. He waters the mountains from his upper chambers; the land is satisfied by the fruit of his work. He makes grass grow for the cattle, and plants for people to cultivate, bringing forth food from the earth, wind that gladdens human hearts, oil to make their faces shine, and bread that sustains their hearts. The trees of the Lord are well watered, the cedars of Lebanon that he planted. There the birds make their nests; the stork has its home in the junipers. The high mountains belong to the wild goats; the crags are a refuge for the hyrax. He made the moon to mark the seasons, and the sun knows when to go down. You bring darkness, it becomes night, and all the beasts of the forest prowl. The lions roar for their prey and seek their food from God. The sun rises, and they steal away; they return and lie down in their dens. Then people go out to their work, to their labor until evening. How many are your works, Lord! In wisdom you made them all; the earth is full of your creatures. There is the sea, vast and spacious, teeming with creatures beyond number—living things both large and small. There the ships go to and fro, and Leviathan, which you formed to frolic there. All creatures look to you to give them their food at the proper time. When you give it to them, they gather it up; when you open your hand, they are satisfied with good things. When you hide your face, they are terrified; when you take away their breath, they die and return to the dust. When you send your Spirit, they are created, and you renew the face of the ground. May the glory of the Lord endure forever; may the Lord rejoice in his works—he who looks at the earth, and it trembles, who touches the mountains, and they smoke. I will sing to the Lord all my life; I will sing praise to my God as long as I live. May my meditation be pleasing to him, as I rejoice in the Lord. But may sinners vanish from the earth and the wicked be no more. Praise the Lord, my soul. Praise the Lord. (Psalm 104)

From these verses one recognizes references to Creation, natural springs, wetland habitat, soil and plant productivity, native plants for animals and food/provisions for all people, the seasons, predator-prey relationships, the work of the steward, the magnificence of God's handiwork, natural renewable resources, natural fire, and humanity as just a minuscule part of the natural system that we are so privileged to be a part of in our daily lives.

In Christian scripture, some key takeaways include the role of Almighty God in Creation, the defining of stewardship practices or actions that landowners continue to apply today, our connection and relationship to nature, and how stewardship is a continuing process during our time on earth. Spiritual responsibility serves to provide the passion and interest in land stewardship.

Catholicism

Many of the concepts presented in the previous sections also apply within Catholic doctrine. The Catholic understanding for care of the land and responsibility for the earth is grounded in Catholicism's Theology of Creation, a set of teachings developed from biblical interpretations, particularly of the Creation stories in the book of Genesis, and the theological tenets of grace and nature. These teachings are presented specifically to challenge the faithful to care for the earth in Pope Francis's recent encyclical *Laudato Si': On Care for Our Common Home* (Pope Francis 2015).

> The creation accounts in the Book of Genesis contain, in their own symbolic and narrative language, profound teachings about human existence and its historical reality. They suggest that human life is grounded in three fundamental and closely intertwined relationships: with God, with our neighbor, and with the earth itself. According to the Bible, these three vital relationships have been broken, both outwardly and within us. This rupture is sin. The harmony between the Creator, humanity and creation as a whole was disrupted by our presuming to take the place of God and refusing to acknowledge our creaturely limitations. This in turn distorted our mandate to "have dominion" over the earth (Genesis 1:28), to "till it and keep it" (Genesis 2:15). As a result, the originally harmonious relationship between human beings and nature became conflictual (Genesis 3:17–19). It is significant that the harmony which Saint Francis of Assisi experienced with all creatures was seen as a healing of that rupture. Saint Bonaventure held that, through universal reconciliation with every creature, Saint Francis in some way returned to the state of original innocence.

For decades, theologians and Bible scholars have clashed over how to interpret Genesis 1:28, in which God instructs Adam and Eve to "fill the earth and subdue it" and "have dominion over the fish of the sea and over the birds of the air and over every living thing that moves upon the earth."

The word *dominion* remains a point of contention. In his interpretation of this passage, Saint Francis held the belief that God calls on humans to be stewards of the earth:

> We are not God. The earth was here before us and it has been given to us. This allows us to respond to the charge that Judeo-Christian thinking, on the basis of the Genesis account which grants man "dominion" over the earth (Genesis 1:28), has encouraged the unbridled exploitation of nature by painting him as domineering and destructive by nature. This is not a correct interpretation of the Bible as understood by the Church. Although it is true that we Christians have at times incorrectly interpreted the Scriptures, nowadays we must forcefully reject the notion that our being created in God's image and given dominion over the earth justifies absolute domination over other creatures.

The views expressed thus far are most often held by those in the Christian faith; however, other religious views that follow are also presented to provide broad context and to acknowledge a common "caretaker" theme central to this subject.

Native American Culture and Spirituality

At first glance, the concept of *stewardship* is apparently at odds with the holistic view of Native American culture and spirituality. The dynamic of subject and object, in which a steward (subject) makes efforts to manage a specific entity or body (object) for the benefit of the steward and/or community, is inconsistent with Native American culture and spirituality, in which the interrelationship between the land, life-forms, and cosmos drives a person's or community's actions. This is a cyclic and interdependent stance, in which individuals/communities do not lord over land or life-forms but work with a sensitivity to the cycles of these lands and life-forms, understanding too that they are part of a larger cycle that maintains a continual balance and harmony. In this way, there is a sacral element to all life that calls on mutual respect, recognizing that different life-forms have needs for their own life requirements. Barbara Thayer-Bacon stated:

> Native Americans' attitudes towards peoples, lands and other life forms such as plants and animals are one of kinship relations. A simple but profound recognition underlies all Indian attitudes towards the organic world. Their relationship with their kin is truly democratic;

> they do not place human beings above animals, plants, stars, or the land, in value or hierarchical status. They are relationships of equity and reciprocity, with an awareness of what other animals sacrifice so that humans may live, knowing that humans will die and their flesh will become soil for plants and plants will be eaten by the animals. Such is the life cycle. (2017, 18)

Chase A. Iron Eyes, a member of the Oglala Sioux, articulates a common stance of all Native American spiritual traditions: The earth is sacred. Whereas European religious and philosophical traditions are predicated on the separation of the sacred and the profane, "our language doesn't even allow us to have that separation" (Jenkins 2020, 151). This stance predates the manifestation of European religious and philosophical systems by several millennia and forms the basis for the indigenous faith-based support of three efforts: care for the environment, sensitivity to the climate and its impact on the environment, and Indigenous people's rights in relation to their care for their land and its resources.

Particularly illustrative of the relation between Native American spiritual beliefs and the practical manifestations of these beliefs are the Blackfeet in Montana. Rosalyn R. LaPier (2017) writes how her grandparents taught her basic, sacred realms of life:

> I learned from my grandparents, both members of the Blackfeet tribe in Montana, about the sacredness of water. They shared that the Blackfeet believed in three separate realms of existence—the Earth, sky and water. The Blackfeet believed that humans, or "*Niitsitapi*," and Earth beings, or "*Ksahkomitapi*," lived in one realm; sky beings, or "*Spomitapi*," lived in another realm; and underwater beings, or "*Soyiitapi*," lived in yet another realm. The Blackfeet viewed all three worlds as sacred because within them lived the divine. The water world, in particular, was held in special regard. The Blackfeet believed that in addition to the divine beings, about which they learned from their stories, there were divine animals, such as the beaver. The divine beaver, who could talk to humans, taught the Blackfeet their most important religious ceremony. The Blackfeet needed this ceremony to reaffirm their relationships with the three separate realms of reality. The *Soyiitapi*, divine water beings, also instructed the Blackfeet to protect their home, the water world. The Blackfeet could not kill or eat anything living in water; they also could not disturb or pollute water. The Blackfeet viewed water as a distinct place—a sacred place. It was the home of divine beings and divine animals who taught the

> Blackfeet religious rituals and moral restrictions on human behavior. It can, in fact, be compared to Mount Sinai of the Old Testament, which was viewed as "holy ground" and where God gave Moses the Ten Commandments.

The inclusive, holistic, cyclic, and sacred qualities of all Creation—in all its realms—are fundamental features of Native American spirituality. These qualities directly influence how the land and its complementary resources—water and air—are managed by Native American practitioners.

Judaism

Judaism, unlike Native American religious concepts and traditions, is dualistic, with clear demarcations between God and the world, between the Uncreated and the created/creatures. God, the Uncreated, created the universe and humanity, set apart from all other creation and atop of that Creation. However, humanity can discern inklings of the nature of the Creator through experience with the created order. Hava Tirosh-Samuelson describes this relationship between the Creator and Creation in this way:

> The more one observes the natural world, the more one comes to revere the Creator, because the natural world manifests the presence of the order and wise design in a world in which nothing is superfluous. Psalm 19:1 expresses this point poetically: "the heavens are telling the glory of God / and the firmament proclaims his handiwork." Psalm 147 (vv. 7–9; 16–18), illustrates how human awareness of the regularity of nature leads to thanksgiving, while according to Psalm 148 (vv. 8–10), all of creation is engaged in praising God and recognizing God's commanding power over nature. Awareness of nature's orderliness, regularity, and beauty, however, never leads the Psalmist to revel in nature for its own sake. In the Psalms, as in the rest of the Jewish tradition, nature is never an end in itself. It always points to the divine Creator, who governs and sustains nature. (2001, 101)

In the Jewish perspective, men and women discern from creation purposeful design and established order and are moved to praise and gratitude for such design and order. Moreover, within this design of creation the Creator demarcates distinctions and boundaries, light from darkness, water from dry land, creatures that crawl on the ground, swim in the sea, and fly in the air. Everything has its place, every creature fulfills a role, and everything has a season (Genesis; Ecclesiastes).

The emphasis on orderliness of Creation explains why in Judaism, we do not find glorification of wilderness and why the cultivated field is the primary model for the created universe in the Bible. Humans are commanded by God to cultivate the earth as a way to preserve and care for what ultimately belongs to God.

The place of humanity, as referred to earlier, is set apart and atop creation. It is the privileged place of the caretaker, created through the single proclamation of God (Genesis 1:27–28), in which God creates mankind "in his image; in the image of God he created them; male and female he created them." God then, revealing God's ultimate design and plan, charges the male and female: "Be fertile and multiply; fill the earth and subdue it. Have dominion over the fish of the sea, the birds of the air, and all the living things that crawl on the earth." By the Middle Ages, Jewish theologians and rabbis debated the meaning of these verses. They made it clear that

> the superiority of human beings over other animals does not entail a license to subdue and exploit. Rather, creation in the image of God entails human responsibility for the whole of creation. Midrash Ecclesiastes Rabba 7:13 expresses human responsibility toward nature as follows: "the Holy Blessed One took the first human and passing before all the trees of the Garden of Eden said: 'See my works, how fine and excellent they are? All that I created, I created for you. Reflect on this and do not corrupt or desolate my world; for if you do, there will be no one to repair it after you.'" This Midrash [commentary] makes clear that humans must neither be indifferent to nature nor bring about its destruction; they must protect nature through their own effort, thereby becoming partners of God, although not cocreators.

This articulation of human responsibility resonates with the widely accepted notion of stewardship as framed by Western culture and thought; humans have a responsibility to take care of the earth as God has deigned them. Humans have a responsibility to maintain the garden, care for its creatures, and continue the processes that God has set forth in the six-day Creation story (Genesis 1:1–31) and the Garden Creation story (Genesis 2:4–25), in which the partnership and intimacy of the male and female are stressed.

Jewish scriptures also reveal a created world as good. But it is not holy, nor perfect, a clear difference from Native American religious sensibilities that experience the earth, the water, and the air as holy. Rather, in Judaism humans are required to perform specified rituals and maintain God's covenant through seasonal observances in order to purge themselves of previous violations and renew their covenant with God. In this manner, the created world is renewed through the justification of humanity. In Judaism, "the earth is the Lord's and all it holds, the world and those who dwell in it" (Psalm 24:1).

Islam

The Qur'an speaks of creation (*khalq*), and it contains 261 verses where this word is used in its various grammatical forms derived from the root *kh l q*. These verses contain references to the human world, to the natural world of the planet from trees to turtles, from fish to fowl, and to the sun, stars, and skies. The very first revelation of the Qur'an to the Messenger of God used this word in its verb form to dramatic effect: "Recite in the name of your Lord who created, created man from clots of blood" (96:1). Creation, the natural world, is the fabric into which the tapestry of life is worked.

Recognition that the world and all that is in it is a result of God's activity is embedded in the Muslim's understanding of land and life. This recognition is core to the Muslim's stance in relation to God; the word *Muslim* means "one who submits" to God; the word *Islam* means "submission." Care for the earth and all that is in it is incumbent upon the Islamic faithful living in humility and honoring and respecting God. Poor stewardship, particularly those who do not respect the care of the land, crops, and its harvest cycles, does not find favor with God.

The Qur'an laid down the basis from which the *Shariah* evolved, which in turn determined the nature of Fiqh (Islamic law) and the subsequent establishment of relevant institutions. The body of the Shariah allows us to deduce three general principles:

1. The elements that compose the natural world are common property.
2. The right to benefit from natural resources is a right held in common.
3. There shall be no damage or infliction of damage, bearing in mind future users.

Hinduism

Hindu thought and practice are diverse and expansive. There is no real single founder; its sacred scriptures span thousands of years without a clear set of authorship. It is a way of life, a philosophy, a devotion, and a means to encounter Ultimate Reality—what adherents of Judaism, Christianity, and Islam may call God. Hindus live throughout the United States, particularly in urban areas, where they can be found excelling in medical research and sciences. Although largely urban in the United States, Hindu thought and practice are directly related to care of the earth and its lands.

The basis for care for the earth, and all life, in Hindu thought and practice can be found in scriptures that strongly center on the value and preeminence of each and every organism in the ecosystem. The Upanishads contain rich teachings, instilling a respect for all life. Indeed, all life relates to and in some way reflects true qualities of Ultimate Reality (divine). For an introduction to the Hindu relation to the care for the land/earth, the Upanishads are most pertinent. The Hindu scriptures also emphasize *bhakti* or devotion to God, and most people of the Hindu faith generally abide by it. In this context, the planet Earth is revered as a mother or goddess, known by such names as *Bhu, Bhumi, Prithvi, Vasudha, Vadundahara,* and *Avni.* She can be seen in temples together with Lord Vishnu and is worshipped as his consort. Similarly, the oldest sacred text in the Hindu corpus, the *Rigveda*, contains inspiring hymns regarding the Earth (Nelson 1998).

The preservation of sacred groves, forests, and gardens as well as pilgrimage to holy sites is not only recommended but also mandated by the Hindu scriptures. Interestingly, some of the sacred groves in the Western Ghats forest regions harbor highly endangered plants and animals even today. Vedic scriptures state clearly that the five major elements of nature—earth, water, fire, space, and air—are sacred. Rivers are particularly revered, and the Puranas teach that a person can gain salvation by bathing in the Ganges.

Buddhism

Buddhism, a religious tradition developed from Hinduism, retains related concepts of interrelatedness of all life and the unique, yet temporary, existence of all life on earth. Gautama Buddha, commonly known as the Buddha, was a *Śramaṇa* (wandering monk) who lived in ancient India. Buddha recognized that local communities could be affected by the behavior of his followers and set rules, for example, that monks and nuns should never relieve themselves in or near running water where people would want to wash or drink. Similarly, he also ruled that monks and nuns should not disrupt the established habitat of any other creature or kill other living creatures, for example, when building new quarters.

Some forms of Chinese and Japanese Buddhism teach the idea of the interrelatedness of everything. This means that humans depend on nature and nature depends on humans. Harming one part of this whole is the same as harming all of it. Therefore, in Buddhist thought, if people learn to live simply and in harmony with the world, the whole of the environment will benefit. For most Buddhists, the guiding principles are to live simply in order to respect all life-forms as well as the balance and peace in nature. Most Buddhists understand and apply these principles to the question of looking after the environment in a variety of ways. Because the Buddha did not have a lot to say about the environment in his day, most teachings about protecting the environment come from modern Buddhist teachers. These interdependent underpinnings of Buddhism are reflected in that Buddhism teaches respect for all life and understanding that all life is interrelated. To disrespect and/or abuse one part of this life endangers one's individual life.

Secular Views of Stewardship

How does this chapter speak to the one who does not embrace religions or some spiritual perspective? The natural world and experience with nature can provide individuals with a deep appreciation and understanding of the vastness and marvels of these "created works." Many landowners, wildlife biologists, foresters, and range managers see this in sunrises and sunsets and through natural processes over time on the land, such as plant succession created by drought, natural fire, and floods, timber harvest, presence or absence of grazing animals, and just the uncertainty and randomness of our natural world. Individuals who are "nonbelievers" are clearly capable of being good land stewards and are motivated in different ways to accomplish care of the land and natural systems. Here we humbly and respectfully acknowledge this fact and do not mean to suggest otherwise.

CASE STUDY: MORMON STEWARDSHIP AND SUSTAINABLE RANCHING
CRAIG D. GALLI

With 670,000 acres of farm and ranchland, the Church of Jesus Christ of Latter-Day Saints (LDS) is one of the largest private landowners in the United States. The Mormon faith puts great emphasis on both environmental stewardship and support of the disadvantaged. This philosophy has led to a tradition of putting their land stewardship philosophy into practice through sustainable ranching and farming operations in Florida and Utah.

LDS Doctrine and Teachings on Environmental Stewardship

LDS theology regarding resource conservation and environmental stewardship revolves around the notion that God created abundant yet finite resources, and humankind would be judged based on the wise and equitable use of earth's resources. In an 1831 statement that LDS church members revere as scripture, church founder Joseph Smith explained:

> For the earth is full, and there is enough and to spare; yea, [God] prepared all things, and has given unto the children of men to be agents unto themselves.

Some have erroneously interpreted this scripture as excusing us from protecting the environment or limiting our consumption. But the next verse states:

> Therefore, if any man shall take of the abundance which I have made, and *impart not his portion*, according to the law of my gospel, *unto the poor and the needy*, he shall, with the wicked, lift up his eyes in hell, being in torment. (Doctrine and Covenants 104:17–18)

We must use natural resources with "judgment" and not "in excess." This mandate imposes a duty to consider the consequences of waste and overconsumption, how to reduce harm to the earth from excessive consumption, while leaving sufficient "abundance" for the disadvantaged.

> I, the Lord . . . make every man accountable, as a steward over earthly blessings, which I have made and prepared for my creatures. (Vs. 13)

Environmental Stewardship and Sustainable Ranching and Farming Practices

A religious person's personal beliefs and the policy statements and doctrines pertaining to environmental stewardship of one's church are less important than the adherent's practical actions to protect the environment. Thomas Carlyle began his hallmark 1840 "Heroes and Hero Worship" lecture with this explanation of religion:

> By religion I do not mean here the church-creed which he professes, the articles of faith which he will sign. . . . But the thing a man *does* practically . . . concerning his vital relations to this mysterious Universe, and his duty and destiny there, that is in all cases the primary thing for him, and creatively determines all the rest. That is his religion.

By that measure, the practices of the LDS Church can be judged favorably, as evidenced by its sustainable building practices and sustainable ranching and farming practices.

Beginning in the 1930s, the church commenced acquiring agricultural lands, originally as a source of food commodities for the church's Welfare Program, which at one time had up to 600 separate farming projects. Since 1980, the church has implemented a strategy to invest many of its reserve funds in farms and ranches. These new agricultural investments were purchased to preserve and enhance the church's reserve funds in a relatively conservative asset while preserving the option of providing food to the needy as disasters or difficulties arose. In 1991, President Gordon B. Hinckley explained the church's investments in agricultural lands:

> The Church does own a number of farm properties. As you know, we have some welfare properties whose produce is used to supply food for the needy. These are operated strictly for charitable purposes and legally qualify for tax-exempt status. Then we have some commercial farm properties. . . . Prudent management requires that this money be put to use. In that process, we have purchased and hold some good, productive farms. They are well operated under capable management, and they yield a conservative rate of return. We have felt that good farms, over a long period, represent a safe investment where the assets of the Church may be preserved and enhanced, while at the same time they are available as an agricultural resource to feed people should there come a time of need.

Over the years as the farm investment program evolved, farm and ranch managers have developed and established a robust model for responsible and sustainable management in keeping with the church's ethos of environmental stewardship. Elder Marcus B. Nash explained in 2014 that the "LDS Church wants to be good stewards of the environment, conserve energy and water resources, minimize pollution, create and maintain cost effective properties and be a good neighbor" (Hinckley 1991). For that reason, the church employs "sustainable agriculture best practices on farms and ranches," such as these:

- Integrated pest management (IPM) for pesticide application
- Use of pheromone traps at orchards as a means of pest mating disruption
- No-till dry farming to reduce erosion and retain moisture
- Good range management practices to avoid overgrazing
- Sponsor of community gardens to give people an opportunity to grow nutritious foods and reduce costs

Two specific examples demonstrate the proactive nature of these environmental management activities.

Deseret Ranches of Florida

Beginning in the 1950s, the church began investing in what is now an approximately 300,000-acre ranch east and south of Orlando, Florida, known as Deseret Ranches of Florida (DRF). The bulk of the lands acquired at that time constituted mostly low-quality, cutover timberland and large swaths of wire grass located between the Kissimmee Chain of Lakes and the St. Johns River. It has taken decades to acquire and improve these ranchlands into viable pastures for cattle grazing, with some pasture development still occurring (figure 16.1). Located on land that borders the St. Johns River for 40 miles, including the headwaters of the Kissimmee River, DRF receives an average of 50 inches of rainfall annually. Thus, water quality protection from nutrient runoff has been critical. To protect and enhance water quality, during the late 1980s and early 1990s,

FIGURE 16.1. Cowboy checking cattle on Deseret Ranch in Florida.

DRF commenced development of a 500-acre retention pond. In 2018, DRF subsequently completed an additional 314-acre retention pond farther down the St. Johns River. Water discharged from these reservoirs is of high quality, generally exceeding the quality of the receiving waters of the St. Johns River floodplain. This retention pond system includes a wetlands complex and riparian areas that provide high-value habitat for migratory birds, alligators, and other Florida species. Named the Jug Island Reservoir Water Treatment System, it was awarded the Florida Commissioner of Agriculture's Environmental Leadership Award for Water Quality.

The DRF employs a full-time, on-site wildlife biologist who helps develop and implement DRF's annual conservation plans and monitors the size and health of wildlife populations. The DRF has documented over 380 wildlife species on-site, including almost 250 species of birds. Protecting and enhancing wildlife habitat have become an important part of DRF's mission. For example, DRF developed and manages a 60-acre wood stork (*Mycteria americana*) rookery consisting of hundreds of island nests of storks and other avian species protected by "deep, moat-like waters and trolling alligators watching for the snakes, raccoons and other animals that prey on the birds' eggs and young." Wading birds and other avian species are most vulnerable during the nesting season. This 60-acre preserve provides important habitat because it offers protection and is strategically located near prime feeding areas. The rookery is particularly important to the wood stork, which the US Fish and Wildlife Service (FWS) listed as endangered in 1984 and reclassified as threatened in 2014. The rookery also provides habitat to other avian species, including roseate spoonbills, great egrets, snowy egrets, and white ibises. The wood stork rookery is a prime example of a DRF project undertaken specifically for its habitat value.

The DRF website summarizes its environmental ethic and conservation efforts as follows (figure 16.2):

> Sustainable resource management has been a key to Deseret's success. The ranch's environmental ethic revolves around principled conservation of wildlife, wetlands and water resources. A trademark characteristic in everything the ranch does is to operate in harmony with the land and natural systems, whether in managing cattle and citrus or in protecting deer, turkey, quail, alligators, cypress trees and wading bird habitat.

The DRF is located in one of the fastest-growing regions of the country, with a projected population of more than 10 million in central Florida by 2080. Urban growth occurring in southeastern Orlando and along the "Space Coast" will inevitably result in development pressure on ranchlands and other underdeveloped areas. For that reason, in 2013, Florida Governor Rick Scott invited DRF to participate in a countywide planning taskforce to examine future urban growth and transportation corridors in Osceola County. One outcome of this long-range planning process was the completion by DRF and Osceola County of the "North Ranch Sector Plan" covering approximately 133,000 acres of DRF property with a planning horizon of 2080. While the North Ranch Sector Plan will include development of some of the DRF lands to accommodate the growing communities, one positive outcome is that ranching operations will continue on tens of thousands of acres of DRF ranchlands for generations to come. And Deseret will remain one of the nation's largest and most productive cattle ranches into the future.

Dale Bills, a spokesperson for Deseret Ranch, explained that when growth does occur,

FIGURE 16.2. A view of cattle grazing through a natural buffer strip between a cypress pond and pastureland.

DRF wants to "create vibrant communities that are environmentally responsible and people-friendly." Some public interest groups voiced opposition to the plan, claiming development near sensitive areas could result in major impacts to wildlife habitat and water resources, while other groups supported the plan. DRF's initial planning approach resonates with the importance of careful land-use planning and environmental stewardship advocated by early and current church leaders.

Utah's Deseret Land & Livestock Ranch

Preservation of native species and habitat enhancement have been of paramount concern at Deseret Land & Livestock (DLL) operations in Utah. DLL is located on over 200,000 acres in the high mountain valleys of northeastern Utah near Woodruff. Elevations range from 6,200 feet to 8,700 feet, making this prime habitat for the greater sage-grouse. In the early 1980s, because of the dwindling population numbers and long before the March 2010 finding by the FWS that listing was warranted under the Endangered Species Act, DLL took proactive measures to protect and manage sage-grouse and its habitat. DLL hired a staff biologist and began studying and improving watershed health and wildlife habitat. After more than 35 years of conservation measures, including sagebrush ecosystem treatments, DLL sage-grouse populations increased compared to nearby populations, while at the same time DLL livestock and elk herds on the ranch increased. Because of these habitat initiatives, in 2011 the Utah Section of the Society for Range Management presented DLL with its 2011 Range Excellence Award "for sustained excellence in improving diverse rangelands and demonstrating profitable livestock production in harmony with healthy wildlife habitats accomplished by a high performing team of managers who openly share their experiences with others."

Over 40% of the sage-grouse's habitat lies on state and private lands. Therefore, efforts by private landowners in undertaking voluntary sage-grouse conservation measures, such as those of DLL, have been critical to the recovery of the sage-grouse, especially where the private property included wet meadows and riparian habitat essential for young chicks to survive. While each private land conservation program differs, federal and state officials work with ranchers and other landowners on long-term agreements to undertake proactive conservation measures that benefit the sage-grouse. According to the FWS, "the future of the sage-grouse depends on the successful implementation of the federal and state management plans and the actions of private landowners." For this reason, DLL continues to cooperate with government agencies and negotiate with both the FWS and the Utah Division of Wildlife Services to establish a Candidate Conservation Agreement with Assurances (CCAA). The effort demonstrates DLL's cooperative attitude and efforts to work with government agencies on a voluntary basis to achieve laudable wildlife protection and habitat objectives using scientifically driven methods beyond what is required by law that at the same time enables a successful and sustainable ranching operation. While the current CCAA has yet to be formally approved by the FWS, DLL continues to abide by the requirements as though it has been.

Conclusion

The leaders and adherents of every faith tradition must decide how to reconcile their life choices with the teachings of their founders and ultimately with the faith tenets they profess to follow. For leaders and members of the Church of Jesus Christ of Latter-day Saints, caring for the community and natural environment has special significance. Joseph Smith taught that the communities the saints built would someday stand as a "standard for the nations" (The Church of Jesus Christ of Latter-day Saints 1838, verses 5–6). As they lived their religion and built up communities, Brigham Young (1868) envisioned that "all the world can say there is a pattern for us, not only in our business and worship, but in our knowledge of [all] things." For Joseph Smith and Brigham Young, the physical design of a community and care for the natural environment encompassed both temporal and spiritual dimensions due to the impact on community members and on future generations. As the LDS Church grows in size and prominence, it will be subject to even greater scrutiny to determine how well its leaders and members live up to these ideals.

Conclusion

We think that a feeling of deep spiritual responsibility as taught and explained by these sacredly held beliefs is a clear and strong internal pathway for many private landowners to feel validated with humility in doing what they personally feel is right in careful management of private lands and publicly owned wildlife that resides therein. We provided an overview of various perspectives that have many common elements. Stewardship is a responsibility given to all of us. The sustainable use and appreciation of these natural resources are an expectation both now and into the future.

CHAPTER 17

Development of a Land Ethic

STEVE NELLE

This chapter articulates some of the philosophical and practical factors that lead to the development of a land ethic in self and in others, building on the thoughts of Aldo Leopold, who wrote extensively about the ethical relationship of humans with the land. Ideas presented here are primarily derived from time spent with many landowners who have demonstrated the true meaning of stewardship.

Introduction: What Is a Land Ethic?

Land ethic is a term that is not used much in our everyday language but is central to any discussions about private land stewardship. All of the following terms are very much overlapping and intertwined, and for practical purposes they are considered to be nearly synonymous: *land ethic*, *conservation ethic*, *stewardship ethic*, and *land stewardship*. Each may have its own colors of meaning, but it is impossible to precisely differentiate these. In this chapter these terms are used interchangeably.

First, it is important to describe what is meant by the word *land* because its meaning here is different and broader than the normal use of the word. According to Aldo Leopold and others who followed him, *land* is an all-encompassing term that includes the soil, water, plants, and animals that form natural or agricultural systems. At a deeper level it includes the water cycle, mineral cycling, plant succession, food chains, food webs, and other ecological processes and dynamics. It is important to note that people are not excluded from this definition; in fact, human influence and management of the land are essential elements of the land complex. In a very real sense, humans are part of the land. Anytime the word *land* is used in this chapter, it has that meaning.

An ethic is a belief—often a moral belief involving a distinction between right and wrong. An ethic is a philosophy of ideas that is held strongly enough that it governs or heavily influences behavior and actions. From this the concept of a code of ethics is derived.

Stewardship involves the responsibility of administering or managing something under one's care. Financial stewardship usually means the wise investment and use of assets to produce benefit and profit while also protecting and growing the principal. Stewardship of natural resources has the similar meaning of using the land for beneficial and productive purposes while at the same time protecting it from degradation or exploitation. Stewardship always looks to the future of whatever is being administered as well as being involved in the day-to-day custodianship of the asset or resource.

The concept of land stewardship comes out very strongly in this excerpt from Aldo Leopold's *Sand County Almanac*: "A land ethic reflects the existence of an ecological conscience, and this in turn reflects a conviction of individual responsibility for the health of land." Note the following elements of the stewardship attitude expressed by Leopold:

- A tender conscience toward the land and its ecological balance.
- The conscience produces inner conviction.
- A strong sense of personal responsibility.
- The result is healthy land with all its benefits.

Putting these ideas together, *land stewardship* can be defined in this way: a deeply held inner conviction that motivates and inspires landowners or land managers to take responsibility to care for and sustain the land entrusted to them. The motives for such stewardship are an important part of the definition. There are typically four primary reasons why land stewards care for the land as they do:

- For care for the land and wanting to do the right thing
- For their own personal benefit, including but not limited to economic benefits
- For the benefit of future generations
- For the benefits accrued to society when land is well cared for

Land stewardship is not primarily about how a person specifically manages a tract of land but is more fundamentally about the inner convictions that guide and energize that management. There is a subtle difference. Land stewardship is really about who a person is on the inside and what the person values, and this in turn dictates and influences how the land is cared for and managed.

Connection with the Land

Before stewardship can happen and before one gains an understanding of the land, there must usually be a strong connection with the land. That connection develops in different ways for different people; the connection can happen anytime in life but often starts in childhood.

If you were raised on a farm or a ranch, you may have gained such a connection with the land, which is the precursor to stewardship. Many of the most well-developed land-people connections exist among this country's agricultural producers who pass those connections down to their children. But not all farmers or ranchers possess the kind of deep land connection that leads to stewardship. We see proof of this in the overgrazed pastures, eroding fields, degraded woodlands, and damaged creeks that still exist in every corner of the country. These symptoms of sick land indicate that there are still many people who live on the land but are not connected to it. The type of connection that is based solely on agricultural production and economic gain is usually not the kind of connection that leads to stewardship.

Perhaps you were raised in the country or on a few acres at the edge of town but not on a real farm or ranch. Living in the country does give one greater opportunity to develop connections with the land, but such connection is not automatic simply because of where you live. There are people who live in the country for many years who never develop a deep connection with the land. They live on the land, but the connection remains shallow and superficial. One may appreciate and enjoy the country lifestyle and the aesthetics of country living without ever developing the kind of connection that leads to stewardship.

For those who were not raised in the country, there can still develop a strong and deep connection to the land. Perhaps a grandparent, relative, or friend lived in the country and you spent time with them growing up and developed a connection. Perhaps you grew up with a love of fishing, hunting, birding, wildflowers, camping, hiking, paddling, or other forms of nature-based enjoyment. These activities can produce a connection with the land that is just as strong as or even stronger than one for someone raised on a farm or ranch. Being raised on a farm or ranch is often viewed as a strong advantage for employment by natural resource organizations, and in some cases, it is an advantage. It helps to know the culture, vocabulary, and mannerisms of agricultural people, and it enables you to fit in better. Knowing something about farming equipment, farm and ranch operations, pesticides, crops, livestock, and so on can help your credibility, especially as a new employee. However, it can also be a disadvantage to have been raised on a traditional farm or ranch. Agriculture is often heavily influenced by tradition, culture, and local customs. Great value is frequently placed on doing things the way that a parent or grandparent did them, even if those were not the best ways. Those agency employees who were raised on poorly operated farms or ranches or where land ethics were not strong are likely to carry that background into their job. Such employees are less likely to notice or address poor management or poor land ethics if that is the context in which they were raised. It should also be noted that certain traditional thinking in the agricultural community can be a very positive thing. Where there is a culture of stewardship and proper care of the land, those traditions are of immense value. There can also be a tradition of cautiously trying new ideas to see how they might be put to good use. Traditional thinking is not bad unless it perpetuates the continuation of poor management.

In sum, it can be an advantage or a disadvantage to have a farming or ranching background depending on whether you were raised with a strong conservation ethic. Likewise, it can be an advantage or disadvantage to have been raised in a nonagricultural setting. Your specific background is not as important as your attitude about the land and connection to the land. If you were not raised on a farm or ranch, it will be important that you spend the time and effort to learn the basics of agricultural practice and production in your area. You do not need to become an expert in agriculture, but you do need to learn enough to converse intelligently and know how agricultural practices affect natural resources.

Leopold summarized the kind of deep connection that leads to a land ethic: "It is inconceivable to me that an ethical relation to the land can exist without love, respect, and admiration for land, and a high regard for its value." In a similar manner, John "Chip" Merrill from a background of ranching and range management describes some of the qualities that accompany a true connection with the land: "One cannot work closely with the land for very long without developing a deep respect and appreciation for its character and capabilities."

Understanding the Land as a Whole

Stewardship of the land cannot usually happen very well apart from a basic understanding of the land. This does not mean that land stewards or their advisors must be technical experts in every facet of the land but that there should be a basic understanding and appreciation of the soil-water-plant-animal-human complex and an innate desire to better comprehend the land and how it works. In the past, our understanding of the land was fragmented by the disciplinary approach. There were soil specialists, agronomists, range conservationists, livestock specialists, foresters, wildlife biologists, fisheries biologists, wetland specialists, hydrologists, and many other specialties. There was frequently little interaction among these; hence the advice given and the management applied were often lopsided and skewed.

In recent years, the concept of holistic management has gained acceptance as the better way to understand and manage the land. It is no longer good enough to cultivate a specialty without also being mindfully aware of how it affects and is influenced by the other parts. Your specialty is but one strand of a complex web. Take every opportunity you can to gain a better ecological understanding of the land as a whole, especially the parts you are weak in. Learning to see the big picture of long-term landscape health is much more important than merely knowing how to kill weeds, plant trees, or produce wildlife. Land stewards and budding land stewards will generally welcome the ideas and assistance of those who are able to understand and articulate the interconnected nature of the land and how each part affects the whole. Baba Dioum, forestry engineer from Senegal said, "We will conserve only what we love; we will love only what we understand; we will understand only what we are taught." This is a good simple explanation of the role of natural resource professionals. Most of us do not directly manage land, but our role is usually to teach landowners and managers to better understand their land and the complexities of ecology and management. As the land complex is better understood, it will be better loved and better cared for. Our role as land advisors is not to prescribe to owners exactly what they should do, but rather, as we patiently teach and owners grow in their holistic understanding of the land, they will increasingly make wise management decisions.

Stewardship Is Not a New Concept

Land stewardship is a popular theme these days. Even for many who do not fully understand what it means, stewardship is considered a positive thing, and people generally want to be perceived as good stewards of the land. Despite the recent buzz of popularity, the concept of land stewardship is not new—it goes back many centuries. The ancient Greek scholar and philosopher Plato wrote about the landscape and management of Greece in the fourth century BC in his dialogue *Critias*:

> In the primitive state of the country, the mountains and hills were covered with soil and there was an abundance of timber. The plains were full of rich earth, bearing an abundance of food for cattle. Moreover, the land reaped the benefit of the annual rainfall; receiving the rainfall into herself and storing it up in the soil. The land let off the water into the hollows which it absorbed from the heights, providing everywhere abundant fountains and rivers. Such was the state of the country, which was cultivated by true husbandmen, who made husbandry their business; with a soil the best in the world and an abundance of water.

Here Plato describes the picture of a healthy landscape with rich soil, good timberland, productive grasslands, abundant water, and a properly functioning water cycle. He then attributes all of this natural abundance to the husbandmen who cultivated the land. We rarely use the term *husbandry* nowadays, but it has essentially the same meaning as stewardship. Plato concluded by stating that they made husbandry of the land their business, which means they exercised knowledge, skill, and dedication in their management and care of the land.

Long before the modern-day stewardship emphasis, some ancient societies understood that responsible care of the land produces abundant and sustainable natural resources and stable, balanced landscapes. Even if *stewardship* has become a somewhat popular and perhaps poorly understood buzzword, it is good that so many people are thinking about it and that it has a very positive connotation. One of our jobs is to help people better understand what stewardship is and what it is not.

What Stewardship Is Not

We have previously defined what land stewardship is, but it is also important to note some of the things that are commonly misunderstood about stewardship, especially since it has to some extent become a modern-day catchphrase. The most common confusion concerning stewardship is that it is primarily about the practices carried out on the land. In the eyes of many, the application of certain practices such as brush and weed control, planting grass, prescribed burning, fence construction, water development, rotational grazing, or food plots is what defines stewardship. These and many other land management practices are merely tools. Tools do not have morality—they can be either good or bad depending on how they are used. If used skillfully, they can produce good results on the land; if used improperly, they cause harm. Using the wrong tool or using the right tool in the wrong way or at the wrong time is often harmful rather than beneficial. Certainly, there is a tie between land stewardship and land management practices, but the two are not synonymous. Stewardship is more about how the practices fit into the overall goal of land health and the skill and finesse in which they are applied for a desirable outcome. Stewardship is what motivates, inspires, and guides the best choice and wise application of practices. Simply completing a checklist of practices is not the same as stewardship.

Another misunderstanding is that the development of a conservation plan, wildlife management plan, ranch plan, forestry plan, or watershed plan must necessarily mean that stewardship is occurring. Such plans, often prepared by agencies or organizations, can be a helpful part of the stewardship process, but too often they are merely paper plans to satisfy bureaucratic mandates and composed of boilerplate cut-and-paste recommendations. In other cases, these plans are prepared hastily without the proper skills, without investing the necessary time and effort, and too often without the active participation of the landowner. A good conservation plan is a living and ever-changing process guided by the landowner's own sense of stewardship ethics and based on his or her own unique land management objectives. Perhaps the best plans are those that reside in the head and heart of stewardship-minded landowners who are constantly thinking about the land and their management and care of it.

Due to the recent rise of stewardship interest across the country, there have arisen hundreds of organized land stewardship initiatives, projects, programs, councils, alliances, and partnerships of many kinds. Certainly, most of these endeavors serve a good purpose and produce positive results. But we should remember that stewardship is not primarily about membership in organizations but is rooted in the relationship between a person and piece of land. One may be a member of many fine organizations, but what is more important is how individuals care for and manage individual tracts of land.

It is also important to note that stewardship is not the hands-off preservationist approach. Some well-meaning environmental advocates think that land should be left alone, letting nature take its course with minimal or no human intervention. That idealistic approach removes the landowner as the custodian and caretaker and removes the production of goods and services needed by society, which in turn generate the economic return needed to sustain private lands.

Yet another misperception is that stewardship is merely an emotion-based response to the land. Although there is a strong emotional aspect of land stewardship, it is not just a warm, fuzzy feeling of affection toward the land. Stewardship involves not only the emotions but also the mind, the hands, and the pocketbook.

Last, land stewardship is not about environmental activism. While there may be a valid place for certain forms of activism to help precipitate cultural change, it tends to divide people rather than bring them together. Landowners and their advisors may have widely differing opinions about current environmental issues and still be firmly unified about the central importance of local land stewardship and land ethics.

Cultivating a Land Ethic in Self

One of the primary purposes of this book is to promote the ethics and practice of responsible land stewardship among our nation's landowners and land managers. The integrity and sustainability of our natural resources depend on a strong sense of stewardship among the owners and operators of farms, ranches, and forests. If you work for a conservation organization, agency, or a related business, one of your jobs is to be able to instill a deeper sense of conservation ethics in those who own, lease, or manage the land. In order to impart a stewardship ethic in others, we must possess it ourselves; we cannot instill in someone else what we do not have in our self. In a sense, we are like the teacher or preacher who must have a real and personal familiarity with the subject before helping others learn it. Therefore, we must each endeavor to cultivate and strengthen the land ethic that we possess to be more effective in our jobs. Just because we have earned a degree in biology, environmental science, range management, agronomy, forestry, wildlife management, or other similar course of study does not mean we have a personal land ethic. It is common, in fact, for a contemporary graduate to gain a good education in the sciences with little or no exposure to the concept of stewardship or land ethics.

If you find that your own land ethic is weak or uninspired, there are ways to strengthen it. Search out those landowners or managers who are known for their stewardship excellence. Get to know them; spend as much time as you can with them; ask them about their operation and why they do what they do. Listen carefully to discover what motivates them to care for the land. If they sense your genuine interest in stewardship, some of them will take you under their wing and become an informal type of mentor. Leasing a tract of land for farming, raising trees, producing livestock, or hunting can be an excellent way to help strengthen your conservation ethic while learning firsthand the many joys and challenges of land management.

CASE STUDY 17.1

Bre Owens

Bre Owens was raised by the land. She grew up on her family's cow-calf operation in Northern California; received degrees from Chico State University in agricultural and wildland range sciences; and did work on both public and private lands in Hawaii, Wyoming, Colorado, and Nevada before returning to work for the California Department of Fish and Wildlife, and later for a private consulting firm. In addition to that work, Bre owns 175 acres of rangeland and pasture in Tehama County, where she runs Dorper sheep and grazes cattle through cooperative grazing agreements on neighboring public lands. Bre says she's a "conservationist at heart," but it's the land and being in production agriculture that ground her stewardship ethic and allow her to be an effective voice for conservation.

> One of the reasons I am so committed to staying engaged in both the conservation conversation and the ranching conversation, and why I feel it's so important to be involved in production agriculture while I'm engaged in those conversations, is there is this element of day-to-day reality that I could not bring to those conversations if I were not involved daily in production agriculture. The only way I know how to elevate the producer voice to an equal footing is to be a producer myself and to have a commitment to that.

FIGURE 17.1. Bre Owens (photo courtesy of Bre Owens).

Hunting, fishing, birding, botany, paddling, backpacking, or other such activities can help cultivate, reinforce, and deepen your relationship to the land. Walk slowly, learn to be observant, indulge your curiosity about the land, and get on your hands and knees to study its complexity. Learning the plants in your area will be one of the primary ways to deepen your appreciation of the land, and many landowners seek out the best "plant people" to advise and assist them.

Natural resource agencies and organizations have a mix of employees with varying degrees of land ethics. In some it is well developed; in others it seems to be nearly absent. There should be no room within the ranks of these groups for employees who lack the strong land ethic needed to do a good job and be effective in assisting landowners. If you find coworkers who seem to lack a strong land ethic, take the time to work with them and help them understand the land better and develop a deeper connection to the land. As employees rise to positions of higher and higher responsibility within organizations, some will be in charge of recruitment and hiring. One of the best things you can do in this role is to try to determine during the interview and selection process which applicants have a true and clear sense of land ethics and what stewardship means. Those applicants who demonstrate such an aptitude will likely be of greater long-term value to the conservation organization or agency.

Promoting a Land Ethic in Others

Some landowners and managers have very deep and well-developed stewardship ethics, while others do not. In large part, this is why we see a big disparity in how people care for the land. When we find those landowners who do already have a strong and clear set of land ethics, we should make the effort to spend time with them and learn from them. This is an important part of our professional development, and it will help us be more effective with other landowners. What we learn from them is often how we deepen our own conservation ethic, and those ideals can be then be passed along to others.

If you are in a position of organizing local landowner events such as field days or seminars, it is good to remember that landowners often learn best from other landowners. These men and women who operate local farms or ranches are often the best spokespersons for stewardship, and their message will usually be well received by other landowners. Find those who demonstrate obvious land ethics and invite them to speak at these events. They may be reluctant to do so, and they may not be a polished speaker, but these are often the most effective presentations to motivate other landowners. Peer-to-peer learning among landowners can also take place apart from formal meetings and presentations. Stewardship-minded landowners are usually generous with their time when asked to share their perspectives and what they are doing. Setting up informal gatherings of just a few people can be an effective and natural way for landowners to exchange ideas about stewardship. Landowners will naturally be careful not to offend or push too hard when it comes to expressing stewardship ethics to fellow landowners.

When you encounter landowners who seem to have little or no interest in the ethics of conservation, it may be best not to invest a great deal of time with them. Perhaps all they want from you or your organization is technical information or financial assistance. Keep the relationship cordial and keep the door open for deeper discussions about stewardship, but spend the majority of your time with people in whom you sense the seeds of a land ethic. Seek out those landowners or managers in whom you see some desire to understand the land and some expression of stewardship, even if it appears rudimentary. Take a genuine interest in them, their place, their goals for the land, their family, and their ideas. Invest the time to get to know them, ask questions, and listen attentively. Observe what is happening on their land, and see if they appear interested in going beyond the basic mechanics of land management. When landowners sense that we are genuinely interested in them and their land, they are more likely to listen, and the relationship will grow. Promoting deeper land ethics in landowners is a challenging endeavor and one that requires time, skill, sensitivity, dedication, and patience. Over time, trust will be built and you may become a valued advisor as well as a friend.

If we can help instill a deeper land ethic among those who own and manage the private lands, the result will ultimately be a gradual trend toward more healthy and productive lands, waters, and wildlife. We do not need to tell land managers exactly what we think they should do with the land; in fact, that is not usually a good approach. But as we develop relationships with landowners whereby we earn their trust and gain their respect, they are likely to listen to our message. That does not mean they will adopt everything we say, but they will listen and consider if and how it might fit into their land management program.

You Are Not the Soloist

As professionals who have been well trained in the science of land management, we like to be the primary source of information and assistance to those we serve. The human ego takes pride in being the principal one that is called for help. As we gain experience, credibility, and a solid reputation, there is a growing tendency to think that we should be able to address the whole range of conservation issues our clients or cooperating landowners face.

However, natural resource management is such a complex discipline that no one has all the answers. No one has the knowledge needed to be the singular authority, and we should never strive to be the exclusive source of expertise among the people we assist. Each of us must recognize that we need others to help us provide well-rounded assistance. We need to develop a large network of fellow professionals whom we can call on when needed. Others have skills and knowledge in certain areas that we do not have, and we have some expertise that they do not have. The goal is not to be the rock star but to help provide the best assistance possible regardless of the source and who gets the most credit. Landowners seem to appreciate when you are willing to refer them to someone else who has the necessary knowledge or experience they need.

Try to diversify and cultivate your professional networks regularly. It should include university scientists, retired ecologists and conservationists, local landowners who excel in certain areas, employees and specialists in other organizations, and anyone who has developed unique land skills. A large professional network will be one of your greatest assets as your career grows. Membership and active participation in professional societies and associations is a good way to get acquainted and cultivate relationships. Take advantage of field days, workshops, and seminars to meet others. Do not be shy about contacting authors and respected experts even if you have never met them in person. Most are happy to help, especially when they see enthusiastic young professionals wanting to excel. As your own skills increase, you will be called on by other professionals to be a part of their network. Getting to know these people and developing your network take some time and effort, but the results are well worth it.

Think of it like being a part of a grand orchestra. There are many skilled musicians, each playing their own part, but no one plays all the parts. The combined sound of many instruments playing their parts far surpasses what even the greatest soloist can do. Some have more prominent roles and some less, but all are vital to the success of a symphony.

The Character of Stewardship

Certain desirable character qualities are often seen in those who possess the ethics of land stewardship. While not always true, genuine land stewards generally have many of these complementary qualities of life that enhance their success on the land. These same characteristics are also important for those of us who work with landowners. These character traits are like a catalyst in a chemical reaction—they help the relationship develop more smoothly and faster than it would otherwise. Following are some of the virtues that landowners respect and that will enhance your success to the degree that you possess these qualities.

Work Ethic

Most landowners, especially those who make their living on the land, are hard workers. They know nothing of the typical 40-hour workweek. Landowners also expect others to be devoted enough to their job to put forth extra effort, above and beyond what is required. You will earn their respect when you demonstrate your willingness to work long, hard hours when needed to get the job done.

Integrity

Land stewards tend to be people of integrity, and it is a cornerstone trait for many of them. This includes integrity with family, neighbors, employees, and hunters, as well as in all business dealings and interactions with the land and livestock. As we exhibit integrity, honesty, and trustworthiness in our dealings with landowners, this will add tremendous value and credibility to our working relationships.

Humility

Humility (literally "of the earth") is an important quality for any profession and is admired by most. Land stewards tend to have a deep sense of humility as they acknowledge how little they actually know about nature and the land. Those of us who work with landowners would do well to cultivate a healthy sense of humility as we acknowledge that we do not have all the answers. A person can be confident and humble at the same time, but it is a challenge to keep them in balance.

Dedication

Most stewardship-minded landowners display a strong commitment and dedication to their work. Likewise, if we are to work effectively with landowners, there must be something deeper than just a job and a paycheck. If you are dedicated to natural resources and to serving landowners, it will be apparent to others and will enhance your effectiveness.

Passion

Enthusiasm and passion are traits that show you enjoy what you are doing and you take it seriously. Many landowners tend to have a quiet, subdued type of enthusiasm, although some are more effusive. A proper expression of enthusiasm backed up by knowledge is contagious and can help propel your message.

Perseverance

Landowners know that there will always be setbacks and difficulties when working with the land. Drought, flood, disease, insects, economic hardships, regulations, family issues, and health problems are some of the adversities that strike farm and ranch people, but the successful ones look beyond the hardship and usually bounce back. Lessons are learned from each setback, which builds even greater resolve and resiliency for the future. As we develop perseverance in dealing with our own setbacks, we will be better able to identify with landowners.

Patience

Patience is a key virtue of life no matter the profession, but for people who work with the land, it is especially important. Nature works on its own timetable, and progress is often slower than we desire. Working with people also requires patience as relationships often develop slowly and gradually as trust and confidence are built. Good things usually take time, and the process cannot be rushed.

Bureaucracy and Stewardship

Inevitably, many of the agencies and organizations that provide services to landowners operate as bureaucracies. This tendency generally stifles the effective and efficient delivery of land management assistance. As employees of these organizations, we must guard against these bureaucratic tendencies if we want to provide the best service. You do not have to become a bureaucrat just because you work for a bureaucracy. Keep the purpose and intent of your work central, and endeavor not to be overly distracted by the multitude of directives from your agency, which may have little or nothing to do with the mission. Focus your attention on providing excellent service to landowners more so than gaining accolades from within your organization.

Naturally, organizations target their most capable employees for promotion to higher positions. This may be what you aspire to, but keep in mind that the higher you climb the ladder, the farther you are from the ground and more detached from working directly with landowners. For many, there is a high degree of personal satisfaction that comes from working closely with landowners, and much of this satisfaction is lost as you are promoted to higher positions. You will sometimes have to choose whom to please and whom to favor the most—your employer or the people you are paid to serve. It can be a difficult balance to attain and maintain.

Stewardship of the Future

Most of the assistance we provide to landowners is the result of a specific request or a current need, and much of our work has a time frame of less than 10 years. However, as our relationships with landowners grow, there may be opportunities to discuss longer-term stewardship issues.

Most landowners, especially agricultural producers, would like for their children and grandchildren to take an interest in the property and to carry on the ownership and management of the land with all the associated benefits of stewarding lands. As we have the opportunity to teach future generations about land principles and land ethics, this is a worthy investment of time that may help build the connection they need to take a deeper long-term interest in stewardship. What is done today and next year on the land is important, but what happens 50 years and beyond is of equal, if not greater importance.

Those who provide conservation assistance to landowners should be familiar with some of the ways in which landowners can ensure that their land stays intact in perpetuity, eliminating the risk of fragmentation. Conservation easements have become an increasingly effective way to do this and are customized to individual landowner needs. Land advisors should become familiar with the land trust organizations of the region and the people who work there, and when appropriate, refer landowners to them.

Conclusion

Taking care of the land is no easy job. It is a complex and ever-changing endeavor with no simple or straightforward solutions or shortcuts. The caretakers and stewards of America's private lands often need and desire the assistance of well-trained, highly motivated professionals who understand the land as well as the people who own and operate the land. Those professionals should have a strong sense of personal land ethics and must be able to inspire stewardship ethics in the landowners with whom they work. Regardless of job title or specialty they must be willing and able to see and understand the big picture of land health and long-term sustainability. Holistic thinking is required rather than the single-discipline approach.

This kind of stewardship assistance is not merely about recommending a generic set of land management practices. Genuine stewardship is the attitude and lifestyle of respect and responsibility toward the land that in turn guide and orchestrate proper and wise management. It requires knowledge, skill, creativity, perseverance, hard work, and patience. The same qualities possessed by stewardship-minded landowners are also needed by the professionals who assist landowners.

The benefits and results of land stewardship are profoundly important to our country and the world. Food, fiber, wood and paper products, wildlife, clean water, carbon absorption, drought mitigation, flood attenuation, and dozens of other natural benefits and services are the result of stewardship. These products and benefits are renewable, sustainable, and cost-effective. Natural resource professionals guided by stewardship principles play an important role in providing these services to society.

SECTION VI

Resources

Glossary of Terms

agrarian. Related to cultivation of land or cultivated land.

Agricultural Conservation Easement Program (ACEP). Program that helps landowners, land trusts, and other entities protect, restore, and enhance wetlands or protect working farms and ranches through conservation easements.

Agricultural Management Assistance (AMA). Program that helps agricultural producers manage financial risk through diversification, marketing, or natural resource conservation practices.

apiculture. The technical term for beekeeping.

appropriative water right. Based on the use of water. Whoever diverts the water first and starts using the water for a beneficial purpose has the right to that water.

Broad Instruction Approach. Directs condemnor in an eminent domain proceeding to examine the fair market value of the property.

Candidate Conservation Agreement (CCA). Voluntary agreement between the US Fish and Wildlife Service and one or more public or private parties to undertake conservation practices to protect or enhance habitat for species identified as candidates for listing by the Endangered Species Act.

Candidate Conservation Agreement with Assurances (CCAA). Expands on the success of traditional CCAs by providing nonfederal landowners with additional incentives for engaging in voluntary proactive conservation through assurances that limit future conservation obligations.

census of agriculture. A complete count of US farms and ranches and the people who operate them, completed every five years.

Civilian Conservation Corps. A voluntary public work relief program that operated from 1933 to 1942 as part of President Franklin D. Roosevelt's New Deal. The CCC provided manual-labor jobs related to the conservation and development of natural resources in rural lands owned by federal, state, and local governments.

common law. Judge-made law.

condemnation. In real estate, refers to the taking of private property by the power of eminent domain to facilitate future public use.

conservation. Refers to the wise use and management of natural resources, with an eye to sustainability and concern that resources not be exploited.

conservation banking. The process of accumulating and selling credits that satisfy regulatory compliance for Sections 7 and 10 of the federal Endangered Species Act and other state and local regulations, for mitigating unavoidable impacts to threatened and endangered species and their habitats and other sensitive habitat areas.

conservation easement. A voluntary, legal agreement that permanently limits uses of land in order to protect its conservation values.

Conservation Reserve Program (CRP). A land conservation program administered by the Farm Service Agency (FSA). In exchange for a yearly rental payment, farmers enrolled in the program agree to remove environmentally sensitive land from agricultural production and plant native species that will improve environmental health and quality.

Conservation Stewardship Program (CSP). A voluntary program of the Natural Resources Conservation Service (NRCS) that provides financial and technical assistance to producers who practice good stewardship on their agricultural lands and incentives for addressing additional resource concerns over a five-year contract period.

contour plowing. An agricultural technique initially adopted popularly during the New Deal era in response to soil erosion that follows topographic contours rather than orienting crop rows up and down a slope.

Convention on International Trade in Endangered Species (CITES). An international treaty to prevent species from becoming endangered or extinct because of international trade. Originally signed by 80 countries in 1975.

Cooperative Endangered Species Conservation Fund (CESCF). Provides funding to states and territories (as authorized under Section 6 of the Endangered Species Act) for species and habitat conservation on nonfederal lands.

development. The process of converting agricultural land to other uses (residential, commercial, industrial, etc.).

Dust Bowl. A period of severe drought and dust storms during the 1930s.

ecosystem services. Refers to the benefits to humans provided by nature (carbon sequestration, water filtration, natural pollination, etc.).

eminent domain. The power to take private or public property interests for a public purpose.

Endangered Species Act (ESA). Signed in 1973 and administered by the US Fish and Wildlife Service, it provides a program for the conservation of threatened and endangered plants and animals and the habitats in which they are found.

Environmental Quality Incentives Program (EQIP). A program of the Natural Resources Conservation Service (NRCS) that provides financial and technical assistance to agricultural producers and nonindustrial forest managers to address natural resource concerns and deliver environmental benefits.

ethic. A belief; often a moral belief involving a distinction between right and wrong. A philosophy of ideas that is held strongly enough that it governs or heavily influences behavior and actions.

Factor-Based Approach. A combination of broad instruction and specific rates approaches in an eminent domain proceeding. States that utilize this approach implement statutes that provide some express terms of property valuation. In addition, if only a portion of the property is condemned, the state will look to see if there was any damage to the property and how characteristics of the property affect the overall value. Some of the factors that may be evaluated are improvements on the property, the presence of growing crops, and the goodwill of business.

fair market value. The price recognized between a willing buyer and a willing seller.

farm. A piece of land focused on producing agricultural products like row crops, hay, livestock, or livestock products.

farm-related income. Includes gross income from farm-related sources received before taxes and expenses from the sales of farm by-products, as well as other sales and services closely related to the principal functions of the farm business.

Federal Land Policy and Management Act (FLPMA) of 1976. Establishes public land policy and guidelines for its administration, as well as provides for the management, protection, development, and enhancement of public lands.

forest. A large area covered primarily with trees.

Forest Legacy Program (FLP). A conservation program administered by the US Forest Service in partnership with state agencies to encourage the protection of privately owned forestlands through conservation easements or land purchases.

Forest Stewardship Program (FSP). A program that works in partnership with state forestry agencies, cooperative extension, and conservation districts to connect private landowners with information and tools needed to manage forests and woodlands and develop forest management plans.

groundwater. The water found underground in the cracks and spaces in soil, sand, and rock.

habitat. The combination of food, water, shelter, and space arranged to meet the needs of wildlife.

Habitat Conservation Plan (HCP). A required part of an application for an Incidental Take Permit under the Endangered Species Act, which allows private entities to undertake projects that might result in "take" of an endangered species.

inholding. Private land that is completely surrounded by public land within the national forest system.

Integrated Natural Resource Management Plan (INRMP). A comprehensive plan created under the Sikes Act that focuses on ecosystem-based management with a goal of managing natural resources to meet stewardship requirements while supporting and enhancing military operations.

joint venture. Since the late 1980s, the USFWS has advanced migratory bird joint ventures where collaborations work to improve wildlife habitat at the landscape scale. As of 2021, there were 18 joint ventures that cross state and international boundaries in North America.

land. An all-encompassing term that includes the soil, water, plants, and animals that form natural or agricultural systems. At a deeper level it includes the water cycle, mineral cycling, plant succession, food chains, food webs, and other ecological processes and dynamics.

Land and Water Conservation Fund (LWCF). A fund established by Congress in 1964 that invests earnings from offshore oil and gas leasing to help "strengthen communities, preserve our history and protect our national endowment of lands and waters." At the state level it provides grants to state and local governments, and on the federal level it is used to acquire lands and other interests.

land ethic. A philosophy of ideas that informs the relationship between humans and nature and governs a land steward's actions related to land conservation and management.

land management. Manipulation of land resources to achieve various objectives.

land stewardship. A deeply held inner conviction that motivates and inspires landowners or land managers to take responsibility to care for and sustain the land entrusted to them.

land-use conversion. A change in land use from an agricultural to a nonagricultural use (e.g., residential or urban).

market-based conservation initiative (MBCI). Program that leverages easements into land protection and agreements between local agricultural operations and military installations. These efforts may include term-limited restrictions on activities on agricultural lands to minimize impacts on military operations. Fair market prices are determined that are acceptable to both the military and local landowners.

mineral interest. All interests belowground (groundwater, oil, gas, lignite, etc.).

mitigation. Refers to the restoration, creation, or enhancement of habitat to compensate for losses in that type of habitat elsewhere.

National Environmental Protection Act (NEPA). The first major environmental law in the United States, which requires federal agencies to assess the environmental effects of proposed major federal actions prior to making decisions.

National Forest Management Act (NFMA) of 1976. Establishes standards for how the Forest Service manages the national forests, requires the development of land management plans for national forests and grasslands, and directs the Forest Service to develop regular reports on the status and trends of the nation's renewable resources on all forest and rangelands.

natural resources. Materials or substances such as minerals, forests, water, and fertile land that occur in nature and can be used for economic gain.

North American Wetland Conservation Act (NAWCA). Provides matching grants to wetlands conservation projects in the United States, Canada, and Mexico to increase bird populations and wetland habitat, while supporting local economies and American traditions such as hunting, fishing, bird-watching, family farming, and cattle ranching.

ownership fragmentation. The breakup of a larger farm or ranch into smaller sections or parcels owned by different individuals or entities.

Partners for Fish and Wildlife Program. Provides free technical and financial assistance to landowners, managers, tribes, corporations, schools, and nonprofits interested in improving wildlife habitat on their land.

prescriptive laws. Define what landowners can and cannot do with their private property.

preservation. Refers to the protection of wildlife and natural resources and elimination of any consumptive use of those resources.

public good. A commodity or service whose benefits are shared by everyone but owned by no one.

ranch. A type of farm that is generally characterized by grasslands, rivers or streams, and woodlands where grazing animals and wildlife coexist.

rangeland. Open country used for grazing or hunting animals.

regulated riparianism. Requires riparian landowners to get a permit that more clearly defines their water right.

renewable energy. Refers to nonconsumptive energy generation from wind and solar resources.

riparian right (surface right). Gives landowners whose land touches a water source, like a river or lake, the right to use water from that source on their land without getting a permit.

Safe Harbor Agreement (SHA). A voluntary agreement involving private or other nonfederal property owners whose actions contribute to the recovery of species listed as endangered or threatened under the Endangered Species Act (ESA). In exchange for actions that contribute to the recovery of listed species on nonfederal lands, participating property owners receive formal assurances from USFWS that if they fulfill the conditions of the SHA, the

USFWS will not require any additional or different management activities by the participants without their consent.

Sikes Act of 1960. Provided for cooperation between the Departments of the Interior and Defense, along with state fish and wildlife agencies, in planning, development, and maintenance of fish and wildlife resources on military installations throughout the United States.

Specific Rates Approach. In an eminent domain proceeding, uses statutes to assess a state's interest in property, valuing certain types of land such as heritage, homestead, and agricultural land. These rates are related to the fair market value of the property, but some types of land receive a premium rate.

State Acres for Wildlife Enhancement (SAFE). A conservation practice within the CRP Program to address state and regional high-priority species in participating states or regions. Producers who voluntarily enroll acres in CRP contracts for 10–15 years receive annual CRP rental payments, incentives, and cost-share assistance for habitat enhancement for sensitive species.

stewardship. The act of caring for something under one's care. In the case of land, the understanding that land will outlive the owner, and therefore management of that land, has impacts beyond individual ownership.

surface interest. All interests aboveground.

surface right (riparian right). Gives landowners whose land touches a water source, like a river or lake, the right to use water from that source on their land without getting a permit.

take. A technical term under the Endangered Species Act, defined as "to harass, harm, pursue, hunt, shoot, wound, kill, trap, capture, or collect, or to attempt to engage in any such conduct."

typology. A classification system based on types or categories.

unified fee ownership theory. The landowner owns everything from the center of the earth to the sky (but cannot interfere with air travel) until one right is sold or given away, at which point that right is "severed."

Voluntary Agricultural Districts (VAD). Localized efforts to maintain agricultural lands by providing relief from some municipal assessments, due process prior to condemnation, and other benefits.

Water Resources Development Act (WRDA). A legislative package that provides for the conservation and development of water and related resources. It authorizes the secretary of the army to conduct studies, construct projects, and research activities that can lead to the improvement of rivers and harbors of the United States.

wetlands. A distinct ecosystem that is flooded by water, either permanently or seasonally.

wildlife. Living things, especially mammals, birds, and fishes, that are neither human nor domesticated.

working lands. Rangelands, farms, and forestlands that are actively managed and used to produce revenue.

Editors

Recognizing national trends in both private land ownership and preparation of professionals with increasingly urban backgrounds, the editors of this book identified an opportunity to create a resource for students, entry-level practitioners, and decision-makers to better understand the importance of private land stewardship and stewards. The editors provided writing support and coordination of authors with the goal of creating a cohesive, thorough, and impactful book that will be relevant to a large audience within the private land community.

Dr. Roel Lopez is director of the Texas A&M Natural Resources Institute and head of the Department of Rangeland, Wildlife, and Fisheries Management at Texas A&M University. He has over 25 years of natural resource and land management experience with academia, federal and state agencies, and private industry. His research focuses on endangered and fragmented wildlife populations, sustainability of military lands, and rural land trends and demographics. Roel received his BS in forestry from Stephen F. Austin State University and his MS and PhD in wildlife and fisheries sciences from Texas A&M University.

Jenny C. Sanders is a project specialist for the Texas A&M Natural Resources Institute. Jenny's work involves building and nurturing partnerships with state and federal agencies, NGOs, and private interests in large-scale conservation efforts. Jenny enjoys serving as a spokesperson for programs she is passionate about and restoring native ecosystems in a way that supports landowner goals. Jenny holds a BS in rangeland ecology and management and an MS in wildlife and fisheries sciences from Texas A&M University.

Dr. James C. Cathey is an associate director for the Texas A&M Natural Resources Institute. Jim is a regional and national expert in wildlife management, Extension outreach and programming, feral hog management, and conservation biology. Jim started hunting at a very early age with his family, and he shaped his passion for the outdoors into his occupation. He holds a BS and MS in wildlife management from Texas A&M University and a PhD from Texas Tech University.

Abigail Holmes Huddleston is a project coordinator at the Texas A&M Natural Resources Institute. She was instrumental in creating resources for the Reversing the Quail Decline Initiative (RQDI), a conservation program that sought to address the decades-long decline of quail populations in Texas through research, education, and outreach. Abigail also provides support and training for County Extension agents, planning and coordination of symposia, and creating reports. Abigail holds a BS in wildlife and fisheries sciences from Texas A&M University.

Senior Advisors

The advisors are national leaders in private and public land stewardship and guided the content, flow, and tone of the book. Their leadership and expertise ensured that the book has national relevance and will serve as an effective tool for educating stakeholders about the critical role private land stewardship plays in the future of our natural resources.

Lesli Allison is the executive director of the Western Landowners Alliance, where she works with private landowners and other stakeholders to advance conservation, sustain working lands, and support rural communities. Previously, Lesli managed a large ranch in the southern San Juan Mountains of Colorado, where she implemented progressive conservation management for 16 years. Lesli holds a BA from Columbia University and an MA from St. John's College, Santa Fe.

Mike Brennan is a program director for the Texas A&M Natural Resources Institute, directing the institute's Wildlife Conservation and Mitigation Program. His work focuses on the development of innovative conservation strategies for the resolution of complex threatened, endangered, and sensitive species conservation challenges nationwide. Mike, who lives in Hoback, Wyoming, has practiced environmental and natural resources law since 1983. Mike obtained his law degree from the University of Oregon School of Law.

Cindy Dohner is an environmental consultant and former regional director for the US Fish and Wildlife Service Southeast Region. Among many other projects, Cindy works closely with the National Alliance of Forest Owners and Conservation Without Conflict to facilitate stakeholder engagement to balance endangered species and climate-related policy with the important role of working forests. Cindy holds a BS in marine biology and biological oceanography from Millersville University and an MS in fisheries from the University of Rhode Island.

Bill Ross is the former secretary of the North Carolina Department of Environment and Natural Resources and now focuses as a lawyer and consultant on projects, cases, and initiatives in environmental law, environmental health, natural resource conservation, military readiness, sustainability, and resilience. Bill holds a BA from Davidson College and a JD from University of Virginia School of Law.

Dr. Neal Wilkins is president and CEO of the East Foundation, which owns and operates about 218,000 acres of South Texas ranchland where scientists and managers work together to address issues important to wildlife management, rangeland health, and ranch productivity. Neal holds a BS in forestry from Stephen F. Austin State University, an MS in wildlife and fisheries Sciences from Texas A&M University, and a PhD in wildlife ecology from the University of Florida.

Contributors

Listed alphabetically, contributors represent broad geographies and diverse disciplines and bring unique perspectives to the content they shared. Some contributors are also editors or advisors and are listed only once.

Daniela Barrientos is a project coordinator at the Texas A&M Natural Resource Institute, where she assists with research projects such as the Texas Landowner Survey and the Texas Water Survey. Daniela holds a BS in environmental sustainability from the University of Texas at San Antonio.

Ty Bartoskewitz is the ranch manager and wildlife biologist at MT7 Ranch outside of Breckenridge, Texas, where he is responsible for planning and supervising ranch operational and capital improvement projects, wildlife and habitat management practices, and outreach and education opportunities. Ty earned a BS in range and wildlife management from Texas Tech University and an MS in wildlife management from Caesar Kleberg Wildlife Research Institute at Texas A&M University–Kingsville.

Bruce Beard is the associate director for policy at the Texas A&M Natural Resources Institute and leads the institute's Military Land Sustainability Program (MLS). The MLS Program supports test and training activities on US Department of Defense (DoD) lands through land management and regional planning, research and policy innovations, and education and outreach. Bruce holds a BS in wildlife and fisheries sciences from Texas A&M University and a master of public affairs in environmental policy from Indiana University's School of Public and Environmental Affairs.

Dr. Drew Bennett is the Whitney MacMillan Professor of Practice of Private Lands Stewardship in the Haub School of Environment and Natural Resources at the University of Wyoming. Drew's work focuses on strategies to balance agricultural production and the conservation of wildlife and other natural resources on private lands in the American West. Drew holds a BS in environmental science and an MBA from University of Denver, an MA in geography from the University of Texas, and a PhD in geography from Oregon State University.

Karrigan Bork is an acting professor of law at UC Davis, whose publications run the gamut from the definitive text on the history and application of California Fish and Game Code Section 5937 to a hatchery and genetic management plan for spring-run Chinook salmon. Karrigan graduated with Distinction and Pro Bono Distinction from Stanford Law School and completed his PhD dissertation in ecology at UC Davis.

Jim Bradbury is an attorney and adjunct professor at Texas A&M Law School, teaching ag law. Jim's practice focuses on land, environmental, water, and regulatory matters, including the imminent water policy issues facing rural Texas. He regularly speaks on landowner rights and water issues as they affect Texas public policy and serves on the board of the Texas Land Trust Council and the Policy Committee of Texas Ag Land Trust, working to conserve working farms and ranches. Jim holds a BS in agricultural economics from Texas A&M University and a JD from University of Idaho College of Law.

Ken Cearley retired as an Extension wildlife program specialist for Texas A&M AgriLife Extension Service based in Amarillo. He is the founder and president of Cornerstone Ranching Solutions®, LLC: A Family Ranch and Wildlife Consulting Team. Previously he has worked as rancher, manager, and consulting wildlife biologist in numerous Texas regions. His family's ranching activities centered in the Hill Country. Most recently he has held stewardship positions with Texas Agricultural Land Trust. Ken holds a BS in wildlife science from Texas A&M University and an MS in range animal science from Sul Ross State University.

Lucas Cooksey is an associate director for the Texas A&M Natural Resources Institute and supports blending policy, planning, and practice into seamless programs that both sustain and enhance active land use alongside meaningful natural resource conservation. His career includes appointments with the US Army Corps of Engineers, the US Army Environmental Command, and the US Army Installation Management Command Headquarters, where he led the conservation programs on over 75 army installations throughout the United States and abroad. Lucas holds a BS in range and wildlife management from Texas A&M University and an MS from Texas State University–San Marcos.

Matthew Crawford is a geospatial specialist for the Texas A&M Natural Resources Institute and Texas Water Resources Institute. As part of the Geospatial Analytics Team, Matt provides support and expertise related to spatial technologies to the Texas A&M AgriLife Research and Extension programs associated with the institutes. He supports the goals of the institutes by assisting staff with maps for presentations, project deliverables, and other products. Matt holds a BS and MS in range science from Texas Tech University.

Chad Ellis is the CEO of the Texas Agricultural Land Trust (TALT), which works to conserve Texas' heritage of agricultural lands, wildlife habitats, and natural resources through the use of working lands conservation easements. Prior to joining TALT, Chad worked for the Natural Resources Conservation Service and the Samuel Roberts Noble Foundation and is active in the development of ecosystem service markets nationwide. Chad holds a BS in natural resource management from Sul Ross State University and an MS in animal science from Angelo State University.

Dr. Andrea Feldpausch-Parker is an associate professor of environmental communication in the Department of Environmental Studies at the State University of New York (SUNY) College of Environmental Science and Forestry. Her research is interdisciplinary in nature, focusing on science and environmental communication with interests in conservation of natural resources through communication and cooperative learning among stakeholders. Andrea holds a BS in fisheries and wildlife from Michigan State University and an MS and PhD in wildlife and fisheries sciences from Texas A&M University.

Craig D. Galli practices environmental and natural resources law in Salt Lake City and has taught environmental law courses at Brigham Young University and the University of Utah. Craig holds a BA and MA from Brigham Young University and a JD from Columbia University.

Wendell Gilgert retired as the Working Lands Program director for Point Blue Conservation Science in California. Previously, Wendell served as a soil scientist in Nevada; soil conservationist, district conservationist, and area conservationist in Northern California; and concluded his 34-year NRCS career by serving as the West Regional wildlife biologist for the NRCS Technical Support Center in Portland, Oregon. Wendell holds a BS in biological science and an MS in plant and soil science from California State University, Chico.

Dr. Jeff Goodwin is the director of the Center for Grazinglands and Ranch Management within the Department of Rangeland, Wildlife and Fisheries Management at Texas A&M University. He previously worked with the US Department of Agriculture's Natural Resources Conservation Service and the Noble Research Institute. Jeff's research interests include the economic and ecological impacts of grazing strategies, sustainable grazing management, and ecosystem services. Jeff holds a BS and MS from Tarleton State University and a PhD from Texas A&M University–Kingsville.

Buddy Huffaker has served as executive director of the Aldo Leopold Foundation since 1999. During his tenure the foundation has interpreted and advocated for Leopold's vision of a land ethic by continuing the Leopold family's commitment to ecological restoration and through education and communication efforts to grow the Leopold Community. Buddy's leadership in the conservation movement includes service on state and federal advisory committees and participation in three White House Conferences on conservation and environmental education.

Dr. Myriah Johnson specializes in beef cattle sustainability and has spent time in research and consulting. She strives for the practical application of research to advance the sustainability and land stewardship of the beef industry. Myriah holds a BS from Oklahoma State University and an MS and PhD from Texas A&M University.

Dr. Angelica Lopez is a research scientist at the Texas A&M Natural Resources Institute. Her research interests are in environmental and natural resource attitudes, behaviors and participation constraints, human dimensions of natural resources, rural sociology, and underrepresented populations. Angelica holds a BS in animal science, an MS in agricultural education, and a PhD in recreation, parks, and tourism sciences from Texas A&M University.

Alison Lund is a program coordinator for the Texas A&M Natural Resources Institute. Alison assists in the organization in management of research, monitoring, reporting, and outreach and training efforts associated with various endangered species and land management projects. Before joining NRI in 2013, Alison worked at the Harte Research Institute for Gulf of Mexico Studies in the Fisheries and Ocean Health Lab. Alison holds a BS in renewable natural resources and an MS in natural resources development from Texas A&M University.

Mark Madison has been the national historian for the US Fish and Wildlife Service (USFWS) since 1999. In that role he supervises the USFWS Museum/Archives with 500,000 historic items chronicling the agency's history over the last century and a half. He also teaches environmental history, environmental ethics, environmental policy, and environmental film at the USFWS's

National Conservation Training Center and Shepherd University. Mark has a PhD in the history of science from Harvard University and degrees in biology and history from Macalester College. He has taught at Harvard and the University of Melbourne (Australia). He also spent three years doing tropical reforestation as a Peace Corps volunteer in the Philippines.

Clay Mathis is the head of the Texas A&M University Department of Animal Science, where he leads the faculty and staff to provide teaching, research, and Extension programs in the largest and most diverse animal science department in the country. Clay holds a BS in animal science, MS in the physiology of reproduction from Texas A&M University, and a PhD in ruminant nutrition from Kansas State University.

Tony Mayoral is a North Carolina native and Texas-raised web development professional and graduate of San Antonio College (2020). While interning for the Texas A&M Natural Resources Institute, Tony's data-mining skills, analysis, and organization were instrumental to the development of chapter 6. Tony's current efforts include a combination of web and sales development.

Gene T. Miller is a Certified Wildlife Biologist® and has worked on restoration of native habitat and wildlife populations on public and private lands from the Mid-Atlantic region to the Southern High Plains of Texas for the past 43 years as an employee of North Carolina Wildlife Resources Commission, Texas Parks and Wildlife Department (retired), and National Wild Turkey Federation. Gene holds a BS in wildlife and fisheries sciences from Texas A&M University and served in the US Marine Corps.

Dr. Blake Murden is the CEO of Shield Land Stewardship Group, where he is responsible for the administration and conservation management of the three Shield-Ayres-Bowen family ranches comprising 37,000 acres in Central and West Texas. Blake has more than 30 years of experience on private lands in Texas, Washington, Oregon, and New Zealand. Blake holds a BS in biological sciences from Tarleton State University and an MS and PhD in wildlife and fisheries sciences from Texas A&M University.

Steve Nelle retired from the Natural Resources Conservation Service after 30 years of service as a range conservationist in Texas. Steve now works with ranchers who desire to better understand and manage their land and to become responsible land stewards. He has had the privilege of working with and learning from some of the very best land and wildlife managers in Texas, and this experience helps him provide high-quality assistance to others. Steve holds a BS in range and wildlife management from Texas Tech University.

Todd Nightingale is a project manager for the Texas A&M Natural Resources Institute. Todd supports the Private Land Stewardship and Military Land Sustainability Programs within NRI and provides additional specialized support to the Department of Rangeland, Wildlife and Fisheries Management at Texas A&M. Previously, Todd served 24 years as a professional forester and wildland firefighter with the Texas A&M Forest Service as a district forester and assistant chief regional fire coordinator, where he helped private landowners, communities, industry, and cooperators conserve and protect natural resources. Todd holds a BS in forestry from Texas A&M University.

Dr. Oswald John Nira studies faith, culture, and spiritual formation. He earned his graduate degrees at the Catholic University of America in Washington, DC, studying Catholic lay leadership, popular religion, Mexican American Catholicism, and Asian spirituality. His current interests include the proactive human response to climate change, pro-life issues, and interreligious dialogue. Born in Honolulu, Hawai'i, and raised in South Texas, he has lived and traveled throughout the United States, partly due to his parents' military career.

Dr. Israel Parker is a research scientist at the Texas A&M Natural Resources Institute and adjunct faculty at the State University of New York (SUNY) College of Environmental Science and Forestry. He conducts research on endangered species, habitat management, invasive species management, and wildlife management techniques. Israel holds a BS in fishery and wildlife sciences from New Mexico State University and an MS and PhD in wildlife and fisheries sciences from Texas A&M University.

Dr. Nils Peterson is a professor in the Department of Forestry and Environmental Resources at North Carolina State University. He conducts research at the intersection between human and natural systems. This includes unraveling the drivers of human behavior relevant to wildlife conservation: environmental literacy, household dynamics, conflict, hunting, and democracy. Nils holds a BS and MS in wildlife ecology from Texas A&M University and a PhD in fisheries and wildlife sciences and sociology from Michigan State University.

Garrett Powers is a geospatial specialist with the Texas A&M Natural Resources Institute. Garrett provides geospatial support within the geospatial analysis team for NRI and the Texas Water Resources Institute. His studies are focused on the application of remote-sensing technologies to solve challenges in natural resource conservation. Garrett holds a BS in spatial sciences and an MS in wildlife and fisheries sciences from Texas A&M University.

Simon Roosevelt is a principal of a multifamily investment group based in the United States and abroad. Simon is an avid hunter and conservationist whose interests have involved him in a wide variety of conservation and environmental projects in North America, South America, and Africa. Current projects include the Conservation Roundtable, a forum that brings together the CEOs of American hunting, environmental, and land conservation organizations to discuss and act on larger issues of common concern.

Kevin Skow is a geospatial manager for the Texas A&M Natural Resources Institute and Texas Water Resources Institute. Kevin provides geospatial support and serves as the lead for several projects within the geospatial and information technology team. He conducts GIS analysis and data management and supports web-based mapping applications focused on natural resource issues. He holds a BS in rangeland ecology and management from Texas A&M University.

Addie Smith is a geospatial specialist for the Texas A&M Natural Resources Institute and the Texas Water Resources Institute. She assists in the development and analysis of spatial data and cartographic requests for natural resource geospatial projects within the institutes. Addie holds a BS and MS in wildlife and fisheries sciences from Texas A&M University.

Jamie Sowell is the fire management staff officer for the USDA Forest Service, National Forests and Grasslands in Texas (NFGT). The forest is nationally recognized as an innovative fire and fuels program utilizing partnerships, Shared Stewardship, and Good Neighbor authority to increase capacity to aggressively apply prescribed fire for fuels mitigation and restoration at a landscape scale in the westernmost range of longleaf pine, and within two of the nation's most populous metroplexes (Houston/Dallas–Fort Worth). He leads one of the largest prescribed fire programs in the nation that promotes interagency, intergovernmental, and community cooperation to reduce wildfire risk and restore fire-adapted ecosystems. His passion for working across boundaries to develop solutions to complex challenges is directly linked to the experience of seeing what is possible when everyone has a common vision and works together toward common goals. Jamie has over 35 years' experience as a prescribed fire specialist and wildland firefighter with the USDA Forest Service.

Decky Spiller is the ranch manager for the Kiowa Division of Silver Spur Operating Company, LLC, in Elbert, Colorado. His role covers daily operations and management as well as long-term strategic planning on 90,000 acres of native, central Colorado rangeland and 1,200 acres of irrigated and dryland hay production. Decky has 30 years of working experience in beef cattle production, beef genetics and breeding, agricultural business, and range management in Texas, Colorado, New Mexico, Wyoming, Nebraska, and Mexico. Decky has a BS in animal science from Texas A&M University.

Temple Stoellinger is an assistant professor at the University of Wyoming and co-director of the Center for Law and Energy Resources in the Rockies. Before joining the University of Wyoming, he served as a natural resource advisor to Wyoming Governor Dave Freudenthal, worked for Shell International B.V., and served as natural resource counsel for the Wyoming County Commissioners Association. Temple holds a BS in environment and natural resources communications and a JD in law from University of Wyoming.

Dr. Matt Wagner retired from the Texas Parks and Wildlife Department as the deputy director for the Wildlife Division. Matt now teaches wildlife law and policy at Texas State University, is an adjunct faculty member at Texas A&M University, and has worked as a consultant for not-for-profits and private ranches in Central Texas. Matt holds a BS in rangeland ecology and management from Texas A&M University, an MS in range animal science from Montana State University, and a PhD in regional planning from Texas A&M University.

Literature Cited

Chapter 1

Isenberg, A. C., and T. Richards Jr., eds. 2017. Alternative wests: Rethinking manifest destiny. Special issue, Pacific Historical Review 86 (1): 4–7.
Linklater, A. 2002. Measuring America: How an untamed wilderness shaped the United States and fulfilled the promise of democracy. New York: Walker.
Madison, J. 1792. Property. Papers 14:266–68.
Malone, L. A. 1993. Reflections on the Jeffersonian ideal of an agrarian democracy and the emergence of an agricultural and environmental ethic in the 1990 Farm Bill. Faculty Publications 597. Williamsburg: William and Mary Law School.
Sioussat, G. L. 1920. Andrew Johnson and the early phases of the Homestead Bill. Tennessee Historical Magazine 6 (2): 14–45.
US Department of the Interior, Bureau of Land Management. 2020. Public land statistics. Denver, CO: National Operations Center. https://www.blm.gov/about/data/public-land-statistics.

Chapter 2

Greer, A. 2012. Commons and enclosure in the colonization of North America. American Historical Review 117:365–86. 10.1086/ahr.117.2.365.
Laurence, E. 1731. The duty and office of a land steward represented under several plain and distinct articles. London: J. and J. Knapton et al.
Reynolds, G. C. 2007. A Native American land ethic. Natural Resources & Environment 21 (3): 16–20. http://www.jstor.org/stable/40924823.
Roosevelt, T. 1909. Address at the conference of governors. In Proceedings of a Conference of Governors in the White House, May 13–15, 1908, 3–13. Washington, DC: Government Printing Office.
Thornton, R. 2005. Native American demographic and tribal survival into the twenty-first century. American Studies 46 (3/4): 23–38. http://www.jstor.org/stable/40643888.
Wilson, J. M. 1847. The rural cyclopedia: Or a general dictionary of agriculture, and of the arts, sciences, instruments, and practice, necessary to the farmer, stockfarmer, gardener, forester, landsteward, farrier, &c. Edinburgh: A. Fullarton.

Chapter 3

Chabbi, A., J. Lehmann, P. Ciais, et al. 2017. Aligning agriculture and climate policy. Nature Climate Change 7:307–9.

Conant, R. 2012. Grassland soil organic carbon stocks: Status, opportunities, vulnerability. In R. Lal, K. Lorenz, R. F. Hüttl, B. U. Schneider, and J. Braun, eds., Recarbonization of the biosphere, 275–302. Dordrecht, Netherlands: Springer.

Cotes, D., P. L. Pert, J. Barron, et al. 2013. Water-related ecosystem services and food security. In E. Boelee, ed., Managing water and agroecosystems for food security, vol. 10, 29–41. Oxford, UK: CABI.

De Deyn, G. B., J. H. C. Cornelissen, and R. D. Bardgett. 2008. Plant functional traits and soil carbon sequestration in contrasting biomes. Ecology Letters 11:516–31.

Dignac, M. F., D. Derrien, P. Barre, et al. 2017. Increasing soil carbon storage: Mechanisms, effects of agricultural practices and proxies. A review. Agronomy for Sustainable Development 37 (2): 1–27.

Domke, Grant M., Brian F. Walters, David J. Nowak, et al. 2020. Greenhouse gas emissions and removals from forest land, woodlands, and urban trees in the United States, 1990–2018. Resource Update FS-227. Madison, WI: US Department of Agriculture, Forest Service, Northern Research Station. https://doi.org/10.2737/FSRU-227.

FAO. 2017. Soil organic carbon: The hidden potential. Rome: Food and Agriculture Organization of the United Nations.

Greenwood, P. L. 2021. Review: An overview of beef production from pasture and feedlot globally, as demand for beef and the need for sustainable practices increase. Animal 15:100295.

Lal, R. 2003. Global potential of soil carbon sequestration to mitigate the greenhouse effect. Critical Reviews in Plant Sciences 22 (2): 151–84.

Lal, R. 2016. Beyond COP 21: Potential and challenges of the "4 per Thousand" initiative. Journal of Soil and Water Conservation 71 (1): 20A–25A.

Leopold, A. 1934. Conservation economics. Journal of Forestry 32:537–44. https://doi.org/10.1093/jof/32.5.537.

Millennium Ecosystem Assessment. 2005. Ecosystems and human wellbeing, vol. 5. Washington, DC: Island Press.

Minasny, B., B. P. Malone, A. B. McBratney, et al. 2017. Soil carbon 4 per mille. Geoderma 292:59–86.

Sanderson, J. S., C. Beutler, J. R. Brown, et al. 2020. Cattle, conservation, and carbon in the western Great Plains. Soil and Water Conservation 75 (1): 5A–12A.

Society for Range Management. 2023. Rangeland ecosystem services: Connecting nature and people. https://rangelands.org/content/files/SRM-Ecosystem-Services-Report.pdf.

Texas Ag Land Trust. 2022. Episode 22—are carbon markets right for you? https://on.soundcloud.com/2xkL6.

USDA. Farm Bill overview. 2023. https: //www.usda.gov/farmbill.

USDA NASS. 2019. 2017 Census of agriculture: United States summary and state data, vol. 1. Geographic Area Series, Part 51, AC-17-A-51. Washington, DC: US Department of Agriculture, National Agricultural Statistics Service.
USDA NASS. 2020. Cattle. Washington, DC: US Department of Agriculture, National Agricultural Statistics Service.
USDA NRCS. 2016. Insects and pollinators: Pollinators by the numbers. Washington, DC: US Department of Agriculture, Natural Resources Conservation Service. https://www.nrcs.usda.gov/conservation-basics/natural-resource-concerns/animals/insects-pollinators.
Von Hase, Amrei, and Jan Cassin. 2018. Theory and practice of "stacking" and "bundling" ecosystem goods and services: A resource paper. Business and Biodiversity Offsets Programme (BBOP). Washington, DC: Forest Trends.
Wilcox, B. P., D. Le Maitre, E. Jobbagy, L. Wang, and D. D. Breshears. 2017. In D. D. Briske, ed., Rangeland systems: Processes, management and challenges, 85–129. London: Springer Nature.
Willard, B. 2022. "Sustainability" vs "ESG." Sustainability Advantage. https://sustainabilityadvantage.com/2022/02/22/sustainability-vs-esg/.
Wohl, E., K. Dwire, N. Sutfin, L. Polvi, and R. Bazan. 2012. Mechanisms of carbon storage in mountainous headwater rivers. Nature Communications 3 (1): 1–8.

Chapters 4 and 5

Conservation Science Partners and American Farmland Trust. 2020a. Conversion of agricultural land to urban and highly-developed and low-density residential between 2001 and 2016 from the Farms Under Threat: State of the States Project, version 2.0. Truckee, CA: Conservation Science Partners.
Conservation Science Partners and American Farmland Trust. 2020b. Land cover and use in 2016 from the Farms Under Threat: State of the States Project, version 2.0. Truckee, CA: Conservation Science Partners.
Conservation Science Partners and American Farmland Trust. 2020c. Nationally significant agricultural land in 2016 from the Farms Under Threat: State of the States Project, version 2.0. Truckee, CA: Conservation Science Partners.
Conservation Science Partners and American Farmland Trust. 2020d. Productivity, versatility, and resiliency in 2016 from the Farms Under Threat: State of the States Project, version 2.0. Truckee, CA: Conservation Science Partners.
Conservation Science Partners and American Farmland Trust. 2020e. State's best agricultural land in 2016 from the Farms Under Threat: State of the States Project, version 2.0. Truckee, CA: Conservation Science Partners.
Sass, E. M., B. J. Butler, and M. A. Markowski-Lindsay. 2020. Forest ownership in the conterminous United States circa 2017: Distribution of eight ownership types—geospatial dataset. Fort Collins, CO: Forest Service Research Data Archive. https://doi.org/10.2737/RDS-2020-0044.
US Census Bureau. 2021. North American industry classification system. https://www.census.gov/naics/.
US Department of Agriculture, Forest Service. 2020. Forest atlas of the United States. https://forest-atlas.fs.fed.us/benefits-forest-industry.html.

US Department of Agriculture, National Agricultural Statistical Service. 2020a. Census of agriculture. https://www.nass.usda.gov/AgCensus/index.php.
US Department of Agriculture, National Agricultural Statistical Service. 2020b. Census of agriculture database. https://www.quickstats.nass.usda.gov.
US Department of the Interior, Bureau of Land Management (BLM). 2020. National timeline. https://www.blm.gov/about/history/timeline#:~: text=With%20 historical%20roots%20dating%20back,established%20in%201812)%20and%20 the.
US Geological Survey, Gap Analysis Project (GAP). 2018. Protected areas database of the United States (PADUS), version 2.0. https://doi.org/10.5066/P955KPLE.

Chapter 6

Bigelow, D., A. Borchers, and T. Hubbs. 2016. U.S. farmland ownership, tenure, and transfer. USDA Economic Research Service, EIB-161. https://www.ers.usda.gov/publications/pub-details?pubid=74675.
Carolan, M. 2018. Lands changing hands: Experiences of succession and farm (knowledge) acquisition among first-generation, multigenerational, and aspiring farmers. Land Use Policy 79:179–89.
Fairchild, E., J. Ulrich-Schad, P. Petrzelka, and Z. Ma. 2022. The lay of the land: What we know about nonoperating agricultural and absentee forest landowners in the United States and Europe. Journal of Environmental Management 313:114991.
Johnston, Robert J. 2018. Ecosystem services. Encyclopedia Britannica, January 5. https://www.britannica.com/science/ecosystem-services.
Lopez, A., J. Dreibelbis, K. B. Hays, et al. 2023. Texas landowner survey. College Station: Texas A&M Natural Resources Institute.
Pew Research Center. 2021. Some digital divides persist between rural, urban and suburban America. https://www.pewresearch.org/short-reads/2021/08/19/some-digital-divides-persist-between-rural-urban-and-suburban-america/.
Snyder, S., Z. Ma, K. Floress, and M. Clarke. 2020. Relationships between absenteeism, conservation group membership, and land management among family forest owners. Land Use Policy 91:104407.
Sorice, M. G., K. Rajala, and U. P. Kreuter. 2018. Understanding management decisions of absentee landowners: More than just presence-absence. Rangeland Ecology and Management 71 (2): 159–62.
United States Census Bureau. 2010. US census 2010. DEC 115th Congressional District Summary File. https://www.census.gov/data/datasets/2010/dec/summary-file-1.html.
USDA. 2020. 2017 Census of agriculture highlights: Producers with military service, ACH17-22. https://www.nass.usda.gov/Publications/Highlights/2020/census-military-producers.pdf.
USDA. 2023. Farms and land in farms: 2022 summary. https://www.ers.usda.gov/data-products/chart-gallery/gallery/chart-detail/? chartId=58268.
USDA NASS. 2021. 2017 Census of agriculture. https://www.nass.usda.gov/Publications/AgCensus/2017/index.php.

Chapter 7

Balukas, J. A., K. P. Bell, and D. M. Bauer. 2019. Classifying private landowners to improve understanding of management decisions and conservation opportunities in urbanizing forested landscapes. Journal of Environmental Management 232:751–58.

Butler, B. J., M. Tyrrell, G. Feinberg, S. VanManen, L. Wiseman, and S. Wallinger. 2007. Understanding and reaching family forest owners: Lessons from social marketing research. Journal of Forestry 105:348–57.

Fischer, A. P. 2012. Identifying policy target groups with qualitative and quantitative methods: The case of wildfire risk on nonindustrial private forest lands. Forest Policy and Economics 25:62–71.

Goyke, N., P. Dwivedi, S. Hitchner, J. Schelhas, and M. Thomas. 2019. Exploring diversity in forest management: Outlooks of African American family forest landowners for ensuring sustainability of forestry resources in the southern US. Human Ecology 47:263–74.

Kendra, A., and R. B. Hull. 2005. Motivations and behaviors of new forest owners in Virginia. Forest Science 51:142–54.

Khanal, P. N., D. L. Grebner, I. A. Munn, S. C. Grado, R. K. Grala, and J. E. Henderson. 2017. Typology of nonindustrial private forest landowners and forestry behavior: Implications for forest carbon sequestration in the southern US. Small-Scale Forestry 16:419–34.

Kluender, R. A., and T. L. Walkingstick. 2000. Rethinking how nonindustrial landowners view their lands. Southern Journal of Applied Forestry 24:150–58.

Kurtz, W. B., and B. J. Lewis. 1981. Decision-making framework for nonindustrial private forest owners: An application in the Missouri Ozarks. Journal of Forestry 79:285–88.

Leopold, A. 1949. A Sand County almanac. New York: Oxford University Press.

Majumdar, I., L. Teeter, and B. Butler. 2008. Characterizing family forest owners: A cluster analysis approach. Forest Science 54:176–84.

Peterson, T. R. 1986. The will to conservation: A Burkeian analysis of Dust Bowl rhetoric and American farming motives. Southern Speech Communication Journal 52:1–21.

Peterson, T. R. 1990. Jefferson's yeoman farmer as frontier hero: A self-defeating mythic structure. Agriculture and Human Values 7 (1): 9–19.

Peterson, T. R. 1991. Telling the farmers' story: Competing responses to soil conservation rhetoric. Quarterly Journal of Speech 77:289–308.

Peterson, T. R., and C. Horton. 1995. Rooted in the soil: How understanding the perspectives of landowners can enhance the management of environmental disputes. Quarterly Journal of Speech 81:139–66.

Richter, K. J. 2005. Using attitudes and motivations to segment the landowner audience: A typology of family forest owners in the Missouri Ozarks and description of management and information behaviors. PhD diss., University of Missouri, Columbia.

Ross-Davis, A., and S. Broussard. 2007. A typology of family forest owners in north central Indiana. Northern Journal of Applied Forestry 24:282–89.

Sanders, J. C. 2005. Relationships among landowner and land ownership characteristics and participation in conservation programs in Central Texas. Master's thesis, Texas A&M University, College Station.

Schweiss, B. E., and J. Dwyer. 2008. Landowner attitudes and perceptions of forest and wildlife management in rural northern Missouri. In Douglass F. Jacobs and Charles Michler, eds., Proceedings of the 16th Central Hardwood Forest Conference, vol. 24, 297–305. Newtown Square, PA: US Department of Agriculture, Forest Service, Northern Research Station.

Sweikert, L. A., and L. M. Gigliotti. 2019. A values-based private landowner typology to improve grassland conservation initiatives. Society and Natural Resources 32:167–83.

Valencius, C. B. 1969. The health of the country: How American settlers understood themselves and their land. New York: Basic Books.

Williams, K. L. 2000. Farmers and ranchers in Calhoun County, Texas: Their land ethic and their interest in natural tourism. Master's thesis, Texas A&M University, College Station.

Wunderlich, G. 2004. Evolution of the stewardship idea in American country life. Journal of Agricultural and Environmental Ethics 17:77–93.

Chapter 8

Benoit, A., T. Johnston, I. MacLachlan, and D. Ramsey. 2018. Identifying ranching landscape values in the Calgary, Alberta region: Implications for land-use planning. Canadian Geographer 62:212–24.

Byerly, H., A. W. D'Amato, S. Hagenbuch, and B. Fisher. 2019. Social influence and forest habitat conservation: Experimental evidence from Vermont's maple producers. Conservation Science and Practice 1:1–12.

Cook, S. L., and Z. Ma. 2014. Carbon sequestration and private rangelands: Insights from Utah landowners and implications for policy development. Land Use Policy 36:522–32.

Dayer, A. A., S. H. Lutter, K. A. Sesser, C. M. Hickey, and T. Gardali. 2017. Private landowner conservation behavior following participation in voluntary incentive programs: Recommendations to facilitate behavioral persistence. Conservation Letters 11:1–11.

Feldpausch-Parker, Andrea, and T. R. Peterson. 2009. Independent review of landowner perspectives: Environmental Defense Fund's landowner conservation assistance and safe harbor programs for the federally endangered ocelot. Report prepared for the Environmental Defense Fund.

Janecka, J. E., M. E. Tewes, L. Laack, et al. 2014. Loss of genetic diversity among ocelots in the United States during the 20th century linked to human induced population reductions. PLOS ONE 9:89384.

National Conservation Easement Database. 2020. https://www.conservationeasement.us/.

North Carolina Sentinel Landscapes. 2021. North Carolina Sentinel Landscapes: A partnership linking conservation, working lands, and military readiness.

Raleigh: North Carolina State University. https://sentinellandscapes.wordpress.ncsu.edu/.
Olive, A., and J. L. McCune. 2017. Wonder, ignorance and resistance: Landowners and the stewardship of endangered species. Journal of Rural Studies 49:13–22.
Peterson, M. N., T. R. Peterson, A. Lopez, and J. Liu. 2010. Views of private-land stewardship among Latinos on the Texas-Tamaulipas border. Environmental Communication 4:406–21.
Quartuch, M. R., and T. M. Beckley. 2014. Carrots and sticks: New Brunswick and Maine forest landowner perceptions towards incentives and regulations. Environmental Management 53:202–18.
Roth, B. J., and R. A. Howie. 2011. Land-use planning and natural resource rights: The Alberta Land Stewardship Act. Journal of Energy and Natural Resources Law 29:471–98.

Chapter 9

America's Longleaf Restoration Initiative (ALRI). 2021. Longleaf learn. http://www.americaslongleaf.org/resources/longleaf-learn/.
Bean, M. J. 1998. The Endangered Species Act and private land: Four lessons learned from the past quarter century. Environmental Law Reporter 28:10701.
Bean, M. J., and D. S. Wilcove. 1997. The private-land problem. Conservation Biology 11:1–2.
Blackfoot Challenge. 2021. History. https://blackfootchallenge.org/history/.
Brick, P. D., and R. M. Cawley. 1996. A wolf in the garden: The land rights movement and the new environmental debate. Lanham, MD: Rowman and Littlefield.
Ciuzio, E. A., W. L. Hohman, and A. M. Strong. 2013. Opportunities and challenges to implementing bird conservation on private lands. Wildlife Society Bulletin 37:267–77.
Department of the Interior (DOI). 2020a. History of the Department of the Interior. https://www.doi.gov/about/history.
Department of the Interior (DOI). 2020b. Land and Water Conservation Fund. https://www.doi.gov/lwcf.
Ducks Unlimited (DU). 2020. Annual report. https://www.ducks.org/about-ducks-unlimited/ducks-unlimited-financial-information/du-2020-annual-report.
Epanchin-Niell, R., and J. Boyd. 2020. Private-sector conservation under the US Endangered Species Act: A return-on-investment perspective. Frontiers in Ecology and the Environment 18:409–16.
Evans, D. M., J. P. Che-Castaldo, D. Crouse, et al. 2016. Species recovery in the United States: Increasing the effectiveness of the Endangered Species Act. Issues in Ecology 20:1–28.
Henson, P., R. White, and S. P. Thompson. 2018. Improving implementation of the Endangered Species Act: Finding common ground through common sense. Bioscience 68:861–72.

James, S. M. 2002. Bridging the gap between private landowners and conservationists. Conservation Biology 16:269–71.
Knapp, C. N., F. S. Chapin III, and J. O. Cochran. 2015. Ranch owner perceptions and planned actions in response to a proposed Endangered Species Act listing. Rangeland Ecology and Management 68:453–60.
Kohm, K. A., ed. 1991. Balancing on the brink of extinction: Endangered Species Act and lessons for the future. Washington, DC: Island Press.
National Park Service (NPS). 2015. Land acquisition history. https://www.nps.gov/subjects/lwcf/land-acquisition-history.htm.
National Park Service (NPS). 2017. Theodore Roosevelt and conservation. https://www.nps.gov/thro/learn/historyculture/theodore-roosevelt-and-conservation.htm.
Ohio's Country Journal. 2017. Agricultural and literary history at Louis Bromfield's Malabar Farm. https://ocj.com/2017/09/agricultural-and-literary-history-at-louis-bromfields-malabar-farm/.
Peterson, T. R., and C. C. Horton. 1995. Rooted in the soil: How understanding the perspectives of landowners can enhance the management of environmental disputes. Quarterly Journal of Speech 81:139–66.
Ramos, M. G. 2008. Civilian Conservation Corps in Texas. Texas Almanac. https://texasalmanac.com/topics/history/civilian-conservation-corps-texas.
Sorice, M. G., W. Haider, J. R. Conner, and R. B. Ditton. 2011. Incentive structure of and private landowner participation in an endangered species conservation program. Conservation Biology 25:587–96.
US Fish and Wildlife Service (USFWS). 2026. Cooperative Endangered Species Conservation Fund. https://www.fws.gov/program/cooperative-endangered-species-conservation-fund.
USDA Forest Service. 2026. Forest legacy. https://www.fs.usda.gov/managing-land/private-land/forest-legacy.
Wyoming Wildlife and Natural Resource Trust (WWNRT). 2020. Annual report 2005–2021. State of Wyoming. https://wwnrt.wyo.gov/status-report#h.p_ID_38.
Yung, L., and J. M. Belsky. 2007. Private property rights and community goods: Negotiating landowner cooperation amid changing ownership on the Rocky Mountain Front. Society and Natural Resources 20:689–70.

Chapter 10

Leopold, A. 1949. The land ethic. New York: Oxford University Press.
Vincent, C. H., L. A. Hanson, and J. P. Bjelopera. 2014. Federal land ownership: Overview and data. Congressional Research Service 7-5700:R42346.

Chapter 12

Baier, L. E. 2020. Saving species on private lands: Unlocking incentives to conserve wildlife and their habitats. Lanham, MD: Rowman and Littlefield.

Bennett, D. E., and H. Gosnell. 2015. Integrating multiple perspectives on payments for ecosystem services through a social–ecological systems framework. Ecological Economics 116:172–81.
Bennett, D. E., H. Gosnell, S. Lurie, and S. Duncan. 2014. Utility engagement with payments for watershed services in the United States. Ecosystem Services 8:56–64.
Bennett, D. E., L. Pejchar, B. Romero, R. Knight, and J. Berger. 2018. Using practitioner knowledge to expand the toolbox for private lands conservation. Biological Conservation 227:152–59.
Buckley Biggs, N., J. Hafner, F. Mashiri, L. Huntsinger, and E. Lambin. 2021. Payments for ecosystem services within the hybrid governance model: Evaluating policy alignment and complementarity on California rangelands. Ecology and Society 26 (1): 19.
Burns, Christopher, Nigel Key, Sarah Tulman, Allison Borchers, and Jeremy Weber. 2018. Farmland values, land ownership, and returns to farmland, 2000–2016. ERR-245, US Department of Agriculture, Economic Research Service.
Byl, J. P. 2019. Perverse incentives and safe harbors in the endangered species act: Evidence from timber harvests near woodpeckers. Ecological Economics 157:100–108.
Epanchin-Niell, R., and J. Boyd. 2020. Private-sector conservation under the US Endangered Species Act: A return-on-investment perspective. Frontiers in Ecology and the Environment 18 (7): 409–16.
Fargione, J. E., S. Bassett, T. Boucher, et al. 2018. Natural climate solutions for the United States. Science Advances 4 (11): eaat1869.
Gosnell, H., and J. Abrams. 2009. Amenity migration: Diverse conceptualizations of drivers, socioeconomic dimensions, and emerging challenges. GeoJournal 76 (4): 303–22.
Johnsen, R. L., V. Butsic, and L. Huntsinger. 2020. Evaluating ecosystem services: Values and return on investment of conservation easements held by the California Rangeland Trust. Sacramento: California Rangeland Trust. https://rangelandtrust.org/ecosystem-service-study/.
Knight, R. L., J. Thorpe, and L. Hidinger. 2011. Rangelands and ecosystem services: Economic wealth from land health? Rangelands 33 (5): 2–3.
Kreuter, U. P., D. W. Wolfe, K. B. Hays, and J. R. Conner. 2017. Conservation credits—evolution of a market-oriented approach to recovery of species of concern on private land. Rangeland Ecology and Management 70 (3): 264–72.
Leopold, A. 1934. Conservation economics. In S. L. Flader and J. B. Callicott, eds., The river of the mother of God and other essays by Aldo Leopold, 193–202. Madison: University of Wisconsin Press.
Meine, C. D., and R. L. Knight, ed. 2006. The essential Aldo Leopold: Quotations and commentaries. Madison: University of Wisconsin Press.
Muir, J. 1911. My first summer in the Sierra. Boston: Houghton Mifflin.
Naugle, D. E., B. W. Allred, M. O. Jones, D. Twidwell, and J. D. Maestas. 2020. Coproducing science to inform working lands: The next frontier in nature conservation. BioScience 70 (1): 90–96.
Seidl, A., D. Anderson, D. Bennett, A. Greenwell, and M. Menefee. 2017. Investing in Colorado: Colorado's return on investments in conservation easements:

Conservation Easement Tax Credit program and Great Outdoors Colorado. Fort Collins: Colorado State University. https://hdl.handle.net/10217/185392.
Texas Land Trends (TLT). 2014. Status update and trends of Texas rural working lands. Texas A&M Institute of Renewable Natural Resources 1 (1): 1–13. http://txlandtrends.org.
Trust for Public Land (TPL). 2019. Conservation almanac: Tracking investment in protected lands. https://conservationalmanac.org.
Western Landowners Alliance (WLA). 2021. Big idea: Habitat leasing. https://westernlandowners.org/policy/habitat-lease/.

Chapter 13

Leopold, A. 1966. Round River. Oxford: Oxford University Press.
Leopold, A. 1991. The river of the mother of God and other essays by Aldo Leopold. S. L. Flader and J. B. Callicott, eds. Madison: University of Wisconsin Press.
Meine, C. D., and R. L. Knight, ed. 2006. The essential Aldo Leopold: Quotations and commentaries. Madison: University of Wisconsin Press.

Chapter 16

The Church of Jesus Christ of Latter-day Saints. 1838. Doctrine and Covenants, Section 115, verses 5–6. Revelation given in Far West, Missouri.
Deseret Ranches of Florida. 2026. https://www.deseretranches.com/.
Hinckley, G. B. 1991. The state of the church. Ensign, May, 51–54.
Jenkins, Jack. 2020. Creation care. In Jack Jenkins, American prophets: The religious roots of progressive politics and the ongoing fight for the soul of the country, 151. New York: HarperOne.
LaPier, Rosalyn R. 2017. Why is water sacred to Native Americans? The conversation. https://theconversation.com/why-is-water-sacred-to-native-americans-74732.
Leopold, A. 1947. Wherefore wildlife ecology? In S. L. Flader and J. B. Callicott, eds., The river of the mother of God and other essays by Aldo Leopold, 337. Madison: University of Wisconsin Press.
Nelson, Lance E. 1998. Purifying the earthly body of God: Religion and ecology in Hindu India. Albany: State University of New York Press.
Pope Francis. 2015. Laudato si': On care for the common home. Our Sunday Visitor, May 2015.
Stewardship. Merriam-Webster Dictionary. Accessed June 8, 2021. https://www.merriam-webster.com/dictionary/stewardship.
Thayer-Bacon, Barbara J. 2017. Land: First Nations examples. Counterpoints 505:33–61.
Tirosh-Samuelson, Hava. 2001. Nature in the sources of Judaism. Daedalus 130 (4): 99–124.
Young, B. 1868. Sermon delivered in the tabernacle. Journal of Discourses 12:52.

Index